This book is to be returned on or before
the last date stamped

Transfer Processes
in Cohesive
Sediment Systems

Transfer Processes in Cohesive Sediment Systems

Edited by

W. R. PARKER

Department of Engineering Science
Oxford University
Oxford, United Kingdom
Formerly Head of Cohesive Sediment Research Institute of Oceanographic Sciences
Taunton, United Kingdom

and

D. J. J. KINSMAN

Freshwater Biological Association
Cumbria, United Kingdom

PLENUM PRESS • NEW YORK AND LONDON

Library of Congress Cataloging in Publication Data

Main entry under title:

Transfer processes in cohesive sediment systems.

"Proceedings of an Estuarine and Brackish Water Sciences Association informal colloquium, held September 14–17, 1981, at the Freshwater Biological Association Laboratory, Windmere, Cumbria, United Kingdom"—T.p. verso.

Includes bibliographical references and index.

1. Water chemistry—Congresses. 2. Water—Pollution—Congresses. 3. Sedimentation and deposition—Congresses. I. Parker, W. R. II. Kinsman, D. J. J. III. Estuarine and Brackish-water Sciences Association.

GB855.T73 1984   628.1'68   84-3303
ISBN 0-306-41663-8

Proceedings of an Estuarian and Brackish Water Sciences Association Informal Colloquium, held September 14–17, 1981, at the Freshwater Biological Association Laboratory, Windmere, Cumbria, United Kingdom

©1984 Plenum Press, New York
A Division of Plenum Publishing Corporation
233 Spring Street, New York, N.Y. 10013

All rights reserved

No part of this book may be reproduced, stored in a retrieval system, or transmitted, in any form or by any means, electronic, mechanical, photocopying, microfilming, recording, or otherwise, without written permission from the Publisher

Printed in the United States of America

PREFACE

In the transfer of chemical species through the aquatic environment the association with fine sediment particles is often of crucial importance. The nature, permanence and kinetics of this association is, in many cases, unknown yet it is central to any effort concerned with predicting the pathways of pollutant transfer and the fluxes along them. It is often unclear to precisely what surface, if any, a pollutant is attached.

The transfer of species between surfaces and solution may be chemically or biochemically controlled. These processes may take place within a host framework which has a time and space dependent structure. The natural straining of this framework resulting from physical and chemical processes moves porewater and the dissolved species in it. Thus, to adequately predict the natural transfer of species within a cohesive sediment system the physical, biochemical and chemical processes must be dynamically coupled.

This informal colloquium examined topics relevant to the association of pollutant and sediment. These included the mineral particle surfaces and the surrounding ionic associations, methods of describing the particles and the physical, chemical, biochemical and biological processes operating in association with fine sediment substrates. The physical processes of sediment transport were deliberately excluded.

The papers assembled here, subsequently submitted by their authors following a general request by those assembled for publication, are presented to provide some record of the meeting. All papers have been subjected to peer review and our thanks are due to the referees for their many constructive comments and for the gift of their valuable time.

CONTENTS

PART I PARTICLE PROPERTIES AND CHARACTERISATION

A New Perspective of Several Approaches to
   Clay/Electrolyte Studies   1
   V.W. Truesdale, C. Neal, and A.G. Thomas

The Heterogeneous Distribution of Anions
   and Water around a Clay Surface with
   Special Reference to Estuarine Systems 17
   A.G. Thomas, V.W. Truesdale, and C. Neal

Humic Substances and the Surface Properties
   of Iron Oxides in Freshwaters 31
   E. Tipping

Sample Pretreatment and Size Analysis of
   Poorly-Sorted Cohesive Sediment by Sieve
   and Electronic Particle Counter 47
   D.I. Little, M.F. Staggs, and S.S.C. Woodman

Size Distributions of Suspended Material in
   the Surface Waters of an Estuary as
   Measured by Laser Fraunhofer Diffraction 75
   A.J. Bale, A.W. Morris, and R.J.M. Howland

Detecting Compositional Variations in Fine-
   Grained Sediments by Transmission
   Electron Microanalysis 87
   P.J. Kershaw

PART II TRANSFER OF POREWATER AND PARTICLES

Escape of Pore Fluid from Consolidating
   Sediment 109
   G.C. Sills and K. Been

The Influence of Pore-Water Chemistry on
    the Behaviour of Transuranic Elements
    in Marine Sediments   127
  B.R. Harvey

The Incorporation of Radionuclides into Estuarine
    Sediments   143
  B. Heaton and J.A. Hetherington

Distribution, Composition and Sources of
    Polycyclic Aromatic Hydrocarbons in
    Sediments of the River Tamar Catchment
    and Estuary, UK   155
  J.W. Readman, R.F.C. Mantoura, and M.M. Rhead

Water Quality Aspects of Dumping Dredged Silt
    into a Lake   171
  M. Veltman

PART III MICROBIOLOGICAL EFFECTS

Autotrophic Iron-Oxidising Bacteria from the
    River Tamar   197
  F.J. Cameron, E.I. Butler, M.V. Jones, and
    C. Edwards

Nitrogen Cycling Bacteria and Dissolved
    Inorganic Nitrogen in Intertidal
    Estuarine Sediments   215
  N.J.P. Owens and W.D.P. Stewart

List of Participants   231

Index   235

A NEW PERSPECTIVE OF SEVERAL APPROACHES TO CLAY/ELECTROLYTE STUDIES

V.W. Truesdale, C. Neal and A.G. Thomas

Institute of Hydrology, Wallingford, Oxon, UK

INTRODUCTION

   Cohesive sediment behaviour is, in part, determined by physicochemical interactions between clay and the accompanying electrolyte solution (Hatch and Rastall, 1965). For example, the rheological properties of a clay can change markedly when the electrolyte in equilibrium with the clay is changed from one containing monovalent cations to one containing divalent cations. It is important, therefore, that knowledge of the chemistry of clay/electroylte interactions be readily available to workers in a number of diverse fields where sediment behaviour is important.

   This paper places the approaches that have been taken to the chemistry of clay/electrolyte interactions in a common perspective, and by doing so it eliminates an unfortunate omission from the existing literature. In essence the problem is that there has been insufficient discussion of the relative merits of the commonly available models for describing clay/electrolyte systems, eg. Helmholtz, Donnan etc. This has led to misapplication of models, and improper use of the variables which derive from them. Because of the extent of the literature upon this subject the paper offers a foundation, not an exhaustive treatment, and accordingly, only the essentials of each approach are given. The benefits of adopting the common perspective are exemplified when it is shown how an over-dependence upon the Helmholtz model has led to serious errors being made in the most elementary measurement made of the system, that of the cation exchange capacity (CEC) of the clay.

   The clay/electrolyte system is defined as consisting of four components; negatively charged clay platelets, uncharged species (e.g. water), cations and anions. Positive charges upon the clay that would involve anion exchange, are not considered. While it is appreciated that a part of the charge upon some clays is variable, being pH-dependent, in common with most existing modelling studies only the fixed-charge case is examined here. While in the rest of the paper the system must only be considered at equilibrium it is worthwhile to begin by considering the immediate origins of the system's components since some readers will undoubtedly start from this point. Let us examine, therefore, the simple system in which a dry, salt-free clay is added to a mixed electrolyte solution. The negative charges of the dry clay will be counter-balanced by cations since the dry clay has to be electrically neutral. Therefore, the cations of the equilibrium mixture will originate from both the electrolyte and the dry clay, but the anions will come only from the electrolyte. Once the clay and electrolyte solution are mixed the ions from both sources will compete freely for whatever positions are available anywhere in the system. Thus, if at equilibrium, there were some way of isolating the clay directly in a dry, salt-free condition the dry clay would generally be found to possess a different complement of cations from its original one. Of course, if the original dry clay were contaminated with salts, additional cations and anions would be introduced into the exchange system and, once dissolved, these too would compete freely.

Cation Exchange Capacity

The CEC of a clay is defined as the amount of cations equivalent to the negative charge (α) upon unit mass of the clay. In defining the CEC in this way it relates directly to a fundamental property of the clay and is independent of the condition of the solution, eg. ionic strength. (This contrasts with other uses of CEC (van Olphen, (1977)). For charge neutrality, in a system concontaining T_+ and $T_- = \Sigma T_k$ and k is an ion, the sum of the positive charges equals the sum of the negative ones, therefore

$$T_+ = T_- + \alpha$$

and since $\alpha \equiv CEC$,

$$CEC = T_+ - T_-$$

In principle, then, CEC is determinable by measuring the total amounts of cations and anions in the system, using, for example, a multiple extraction technique (Zaytseva, 1962).

A GENERALISED MODEL

In rationalising the several modelling approaches used hitherto it is advantageous to establish a generalised model which fulfils

all the needs of the individual models used in the different
approaches. Of course, this can only be performed with considerable
hindsight, but, making the individual models special cases of the
generalised one facilitates identification of their real differences. In practice, this approach differs from the earlier ones by
placing the minimum of restrictions upon the whereabouts of the
components.

A generalised clay/electrolyte system would allow its ions and
water to reside in a solid (clay) phase and a liquid (bulk solution)
phase and in an unspecified number of compartments within an interfacial region. This approach allows the term, phase, to be applied
rigorously, as being an environment that is not significantly
inhomogeneous (Moore, 1974). Following existing practice, the solid
(clay) phase is considered to contain neither water nor ions which
are involved in the equilibrium, and the clay platelet is thereby
taken solely as a source of localised negative charge. Throughout
this paper any ions or water molecules of the interfacial region
which are close to the clay surface, and are therefore, somehow
distinctly different to those in solution are called proximal
species; this avoidance of the term, adsorbed, which others might
have used, is discussed later. Figure 1 represents the case of the
generalised model for three proximal compartments containing only
cations and water. The figure should satisfy several possible
arrangements of compartments. One extreme case is a layer-upon-
layer arrangement. The other is that in which all four interfacial
compartments contact the platelet surface; in this case each plot
is really a composite of four separate graphs each with its own
origin.

A RATIONALISATION OF EARLIER MODELLING

As a prelude to discussion of the more intricate problems
associated with the literature it is useful to describe the salient
features of the models that are the source of much confusion,
demonstrating how they are special cases of the generalised model.

The Donnan Membrane Model

Classically, the Donnan model (Donnan, 1935) possesses a
membrane which separates two solutions, the inner and outer. The
membrane prevents certain ions (the non-diffusible ones) from
diffusing from the inner to the outer solutions. In this classical
model ions of each solution move freely and thus the solutions form
two separate phases. However, as stated by Donnan (1935) this
situation is not a prerequisite of the model's application. The
model describes the equilibrium concentrations of the diffusible
salts in the system. At equilibrium, the chemical potential of

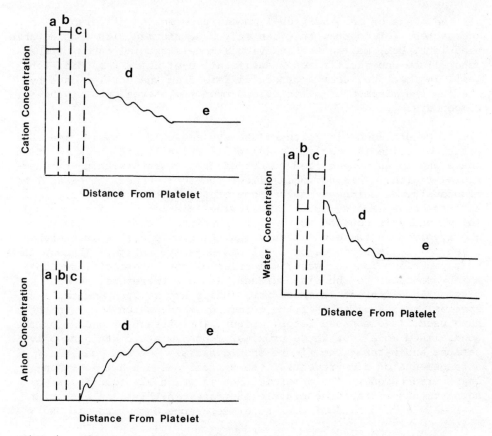

Fig. 1. The generalised model for the case of three proximal layers. Layer a contains only cations, layer b both cations and water, layer c just water; proximal anions are neglected. d, the other part of the interfacial region contains all three components in solution. e represents the bulk solution. The wavy line indicates the absence of a precise species distribution in solution.

any diffusible salt has to be equal on both sides of the membrane (Donnan, 1935). This condition leads to the well-established relationship between the activities of the ions of a given salt in both the inner and outer solutions (Donnan, 1935; Wiklander, 1964; Babcock, 1963). Thus, for an electrolyte of the general type $M_n A_m$,

CLAY/ELECTROLYTE STUDIES

$$\{a_M^n \cdot a_A^m\}_{inner} = \{a_M^n \cdot a_A^m\}_{outer} = K$$

where, a, refers to activity, and K is a constant. When applied to clays the non-diffusible ions are replaced by the fixed negative charges upon the platelets and a surrogate membrane is implied (Babcock, 1963; Overbeek, 1956).

The Donnan model involves no assumption about scale. Thus, insistence upon either a micro (Wiklander, 1964; Kruyt, 1952; Bolt, 1955, 1967) or a macro (Moore, 1974; Gast, 1977; Davis, 1942; Schofield and Talibuddin, 1948; White, 1979) scale application is unnecessary. When applied on a micro-scale the fixed negative-charges will preclude homogeneity within the inner solution, and the inner solution is aptly described as the inter-facial region.

It is important to understand that the Donnan model is not a charge-distribution model, and therefore cannot, by itself, describe the distribution of the ions in the interfacial region. However, in principle, this deficiency can be overcome by incorporating an independently specified charge-distribution by considering the electrochemical potential, $\bar{\mu}_i$, of each ion, i. Thus:

$$\bar{\mu}_i = \mu_i + z F \psi \qquad (1)$$

where μ_i is the chemical potential, z is the valency of the ion, i, F is the Faraday constant and ψ the electrical potential in which the ion, i, is situated. The value of the electrical potential is supplied by the chosen charge distribution. Extending the Donnan model in this way reminds one that the Donnan model, by itself, is only concerned with the distribution of salts (for which the electro-chemical potential is zero), and therefore, can only describe the relative behaviour of ions, for example, as a ratio of their activities. In contrast, the extension deals with the behaviour of a single ion. In effect the applied charge-distribution defines the position of the membrane, and thus, the inner solution's volume; a property the Donnan model, itself, does not possess. When the Donnan model is applied on a macro-scale the inhomogeneity produced by the fixed negative-charges of the clay platelets will be negligible and the inner solution can be taken to be a phase. Even so, without a knowledge of the position of the 'membrane' the usefulness of the model will be severely limited.

The Stern Family of Models

The Stern family of models describes the distribution of charge around a clay platelet, from which a distribution of ions may be

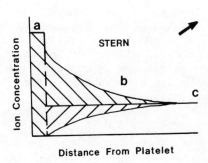
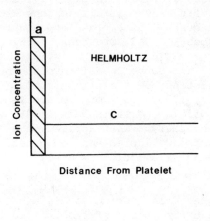
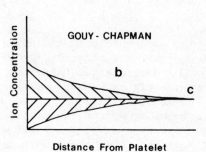

Fig. 2 A rationalised view of the Stern family of models. While a represents the proximal layer of cations, b represents the interfacial solution, and c the bulk solution. The two types of hatched areas identify the cation and anion NICs (water) (see later) and their sum, the CEC.

inferred. A given charge distribution can be satisfied by many ion distributions, and the variants will be distinguishable by equilibrium constants or selectivity coefficients which express the competition between ions for a given position in the system.

The Stern model (Stern, 1924) (Figure 2) has proximal cations (defined earlier in this paper) and the distribution of charge in solution is given by the Poisson-Boltzmann equations (Kruyt, 1952; Shaw, 1970). Equilibrium constants or selectivity coefficients are required to define the ion distribution in the proximal layer (Bolt, 1967). In deriving the ion distribution in solution from the charge distribution it is assumed that the ions are point charges and that the electrochemical potential of each ion (eqn. 1) is constant throughout the solution. The concentration of each ion in solution varies monotonically with distance away from the clay, converging to an asymptote. At some distance from the clay the lack of complete convergence will be masked by the inherent random motion of the ions. The solution beyond this distance is, therefore, the bulk solution and the volume nearer the clay is the interfacial region. In this model, the interfacial region contains both proximal cations and ions in solution. For the purpose of standardisation the approximate position of the boundary between the bulk solution and the interfacial region could be defined arbitrarily as, for example, the distance at which the electrical potential attenuates to a given fraction of its value at the surface of the proximal-cation layer (Wiklander, 1964).

The Helmholtz model (Helmholtz, 1879) (Figure 2) is one extreme form of the Stern model in which there is no net charge upon the combination of clay platelet and proximal cations. As a consequence the whole of the solution constitutes a single phase, the bulk solution. The interfacial region contains only proximal cations. The proportion of each cation in the proximal layer has been defined in a wide variety of ways, and many of these have been reviewed (Bolt, 1967). The Gouy-Chapman model (Chapman 1913; Gouy, 1910) (Figure 2) is the other extreme case of the Stern model, and possessing no proximal ions, has all the ions of the interfacial region in solution, where the Poisson-Boltzmann distribution applies. Note that like the parent model, but unlike the Donnan model, these two derivatives of the Stern model are charge distribution models and describe the whereabouts of the ions precisely. Historically the Stern model was formed by amalgamating the Helmholtz and Gouy-Chapman approaches (Moore 1974; Kruyt, 1952). However, in this rationalisation, with its requirement to classify models according to their general applicability, it is more helpful to take the reverse view where the Gouy-Chapman and Helmholtz models are special cases of the Stern model. Moreover, the historical perspective can hide the fact that the Stern model might, at first sight, appear to give a better fit to experimental data than either of its predecessors, simply because it contains extra parameters (Bard, 1974).

Major Confusions in Earlier Modelling Approaches

Much confusion surrounds the Donnan model and its relationship to other models. Here, three main inconsistencies of its application are identified. Although discussed individually in earlier papers these appear in several different combinations making their identification difficult. One problem is that while some workers (Moore 1974; Bolt and Bruggenwert, 1976; Helmy, 1963) have assumed that the Donnan and Gouy-Chapman models are rival charge-distribution models, and are therefore mutually exclusive, others (Wiklander, 1964; Murthy and Ferrell, 1972; Devine et al., 1973; Berner, 1971; Marshall, 1964) have assumed that they are complementary, and can be applied simultaneously. The primary part of this problem was settled above when it was shown that the Donnan model is not a charge-distribution model. The remaining part, therefore, concerns the advisability of melding the two models. That this is acceptable, although fruitless, can be seen from the fact that the Gouy-Chapman model only differs from the Donnan model in its use of additional assumptions needed to supply a charge distribution. Indeed, the Donnan equilibrium is inherent within the Gouy-Chapman model. Thus, since both are equilibrium models they have to comply with the need for the chemical potential of any dissolved salt to be constant throughout the system. Moreover, both possess a bulk solution and an interfacial region and in the melding the bulk solution of one corresponds with that of the other, and similarly for the interfacial regions. Even so, the melding of these models is potentially misleading since it is liable to obscure the fact that the simplicity and paucity of the assumptions of the Donnan model allow it to function with a wide variety of other charge distributions, even those predicting an apparent discontinuity in concentration at the 'membrane'.

A second problem is that while some workers (Overbeek, 1956; Bolt, 1967; Posner and Quirk, 1964; Davis, 1942, 1945) insist that the Donnan model must have a uniform distribution of charge in its inner solution, others (Donnan, 1935, Marshall, 1964) have deemed this to be unnecessary. The correctness of the latter viewpoint is verified by the acceptability of melding the Donnan with the Gouy-Chapman model to provide a non-uniform charge distribution in the interfacial region of the hybridised model. The incorrect supposition of a uniform charge distribution (Overbeek, 1956; Bolt, 1967; Posner and Quirk, 1964; Davis, 1942, 1945) appears to be an artefact of the original application of the Donnan model to systems containing only free-moving ions, and colloids (Donnan, 1935; Glasstone, 1962).

A third problem is that some workers, in using the Donnan and Gouy-Chapman models simultaneously, have maintained that the Donnan inner-volume is effectively independent of ionic strength (Wiklander, 1964; Devine et al., 1973). That this is incorrect follows from the fact that the hybrid of the two models must satisfy

the assumptions inherent in the component models. Since the volume
of the interfacial region of the Gouy-Chapman model is sensitive to
ionic strength (van Olphen, 1977) the position of the 'membrane' of
the hybrid model must be similarly sensitive.

Ranking the Earlier Models

Having clarified the differences between the distribution
models it would be remiss not to try to rank them. Ideally this
ranking should first entail the quantification of the parameters
of each model, eg. the Donnan inner-solution volume, the Gouy-
Chapman double-layer distance, etc. Subsequently, these quantities
should be judged for their reasonableness and, where possible, for
their consistency with values obtained from other sources, eg.
osmosis studies. Since the Donnan model has not yet been applied
in this manner without concomitant use of Gouy-Chapman theory, the
Stern models currently offer the most useful solutions. Moreover,
notwithstanding that of these the Stern model, itself, will give
the best fit, initially the Stern and Gouy-Chapman models would be
preferred to the Helmholtz model because they, unlike the Helmholtz
model, offer an explanation for the phenomenon of negative adsorption. So-called negative adsorption (see later for clarification
of adsorption) is manifest as the increase in the total concentration of either anions or cations (some earlier workers (Wiklander,
1964) only considered anions) that occurs when a dry, salt-free
clay is added to an electrolyte solution. The Stern and Gouy-
Chapman models offer a quantitative treatment of the phenomenon
based upon the non-uniform distribution of ions in solution
(Schofield and Talibuddin, 1948). The Donnan model is consistent
with the existence of this phenomenon but cannot quantify it
since this requires the 'membrane' position to be defined.

QUANTIFYING THE CLAY/ELECTROLYTE SYSTEM

Unfortunately, so far, direct measurements of the ion distributions around a typical clay platelet as depicted, for example,
by the Gouy-Chapman model are impracticable. Consequently, direct
verification of the above mentioned models is, as yet, impossible.
Moreover, quantification of the clay/electrolyte system, by
separation of the total amount of each species in the system into
that part "belonging" to the clay and that to the solution, can
only be performed <u>arbitrarily</u> (Laudelout et al., 1968; Neal et al.,
1982). There is no <u>absolute</u> means of performing this separation
when any of the clay's negative charge is counter-balanced by ions
in solution; attempts to do so ultimately involve the illogical
step of deciding which of two identical ions in solution in the
interfacial region should 'belong' to the clay, and which to
solution. The argument for arbitrariness is analogous to the

division of a measured EMF for a cell into two half-cell potentials (Glasstone, 1962; Willard et al., 1965). Thus, in both the electrochemical and the clay cases there is no additional physical property that can be used to determine how an <u>absolute</u> separation should be made. (Of course, if clay/electrolyte systems were known to behave according to the Helmholtz model an <u>absolute</u> separation could have been based upon the fact that the charge-balancing cations, the proximal cations, reside on the clay surface and are physically separable from those in solution).

Notional Interfacial Contents

It is important to recognise that in dividing the total amount of a given species arbitrarily between that "belonging" to the clay and that to the solution, conceptually one moves away from the real system to a notional one. Accordingly one abandons the real bulk solution and the real interfacial region, and substitutes instead, notional ones. Nevertheless, to maintain the notional system as close as possible to the real one the notional bulk solution can be assigned the same composition, but of course not extent, as the real bulk solution. Thus, the amount of a given species, i, in the notional bulk solution can be expressed as the product of T_j, the total mass of an arbitrarily chosen reference species j in the system, and $C_{i(j)}$ the concentration of i with respect to j in the real bulk solution. The total amount of species i in the system, T_i, is given by,

$$T_i = NIC_{i(j)} + T_j \cdot C_{i(j)}$$

where $NIC_{i(j)}$ is the notional interfacial content of i with respect to the reference species, j. In this particular notional system the reference species resides entirely in the notional bulk solution. Thus, when the reference species, itself, is considered (i = j),

$$T_j = T_j \cdot C_{j(j)} \quad \text{and} \quad NIC_{j(j)} = 0$$

since $C_{j(j)} = 1$.

Given n species in a system the system's complete description will entail (n-1) NICs; one species being required as reference. With n species there are n sets of (n-1) NICs, since any species can be the reference. This emphasises the <u>relative</u> nature of the description afforded by the NIC concept. A detailed treatment of NICs has been given elsewhere (Neal et al., 1982) with an account of their measurement, use and relationship to other factors, eg. CEC and negative adsorption.

Ion Surface-Excesses

As the concept of ion surface-excesses (Bolt, 1955) resembles the NIC concept it is useful to introduce it now although it is not required until later. The surface-excess, γ_i, of the ion i is given by,

$$\gamma_i = T_i - V C_i \qquad (2)$$

where C_i is the concentration of the species i in the real bulk solution (m equivalents/ml) and V is the liquid content of the system (ml). By analogy with the NIC concept Bolt (1955) can be seen to have been defining a notional bulk solution such that it contained all of the liquid of the system, but had the same composition as the real bulk solution. The main disadvantage of this approach, and hence the reason for its being superseded by the NIC concept, is that it suggests that only one reference species, the liquid, is plausible. It also suffers from the assumption that the liquid volume, V, is determinable. In fact, without a constant water concentration throughout the real system this is not so. Thus, the mass of the water is determinable by, say, evaporation and weighing, but this can only be converted to a volume upon the assumption of a constant density.

SOME PROBLEMS WITH TERMINOLOGY

The Use of "Adsorbed"

The term, adsorbed, has been studiously avoided in this paper because it has already been used for an alarmingly large number of different concepts. Some of the definitions used include, all (Kruyt, 1952; Gast, 1977; Overbeek and Lijklema, 1959) or an arbitrary fraction (Davis, 1945; Kelley, 1948) of the ions in solution in the interfacial regions of the Donnan and Gouy-Chapman models; the proximal cations (van Bladel and Laudelout, 1967; Laudelout et al., 1968; Yong and Warkentin, 1966; DST, 1974) of the Helmholtz model; the proximal cations of the Stern model (the specifically adsorbed ions); those ions calculated as a surface excess in the Gouy-Chapman model (Bolt, 1967; Helmy, 1963); all the ions in the bed-volume of an ion-exchange column (Thomas, 1965) (this is reminiscent of a macro-Donnan model), and all the ions liberated from a clay by a given chemical treatment (Laudelout et al., 1968; Neal, 1977; Laudelout, 1965; Russell, 1970) (ie. empirical definitions). This multitude of definitions presents a problem especially when many writers have not stated a definition, thereby leaving it unclear as to which they are using. Moreover, some workers have used two incompatible definitions simultaneously (Wiklander, 1964). The use of the new terms proximal and NIC circumvents the problems caused by the history of the use of 'adsorption'; whereas proximal involves a division of species according to mechanistic principles, NICs relate the thermodynamic principles (Neal et al., 1982). Notwithstanding our objections to the term adsorbed, the term negative adsorption is still retained but only for the want of a replacement with which to describe the phenomenon succinctly.

The Use of "Exchangeable Cations"

The term exchangeable cations has been used to apportion the CEC between each of the individual cations (Bolt, 1967). However, like the term adsorbed it is afflicted by multiplicity of definition. One definition relies upon the Helmholtz model such that the sum of the proximal cations yields the CEC. Another, more general definition, entails an arbitrary partitioning of the anions of the system amongs the cations present. This is effected using either anion surface excesses (Bolt, 1967) or the total amount of anions (Bolt, 1967) (or of the major anion (Laudelout et al., 1968; Thomas, 1965)) present. The use of the new term NIC circumvents these problems by allowing these two earlier definitions to become special cases of a generalised approach. Thus, in circumstances where the Helmholtz model would be applicable, the cation NICs, defined using either water or any anion as reference species, would be analogous to Helmholtz exchangeable cations. Moreover, Bolt's use of exchangeable cations is provided for by cation NIC (total anion).

SOME PROBLEMS WITH PRACTICAL WORK

Many recipes (Bower et al., 1952; Avery and Bascomb, 1974; USDA, 1969; Reitemeier, 1946; Hesse, 1971; Chapman, 1965; Dewis and Freitas, 1970; Isaac and Kerber, 1971) for CEC are deceptively dependent upon the validity of the Helmholtz model and by de-fault measure, instead of the CEC, something resembling the sum of the cation surface excesses. In effect, this group of recipes overlooks the contribution to the CEC of the anion surface-excesses. The CEC equals the difference between the sum of the cation surface excesses and the sum of the anion surface excesses (Bolt, 1967). It will be recalled that ion surface-excesses are, in reality, indeterminable. Nevertheless, for ease of understanding, this discussion is continued as though ion surface-excesses were determinable; this is not to be taken as advocating their continuance elsewhere. Two major sub-groups of recipes can be identified depending upon whether the magnitude of the anion surface-excesses is governed by the salinity of the sample or that of the reagents.

The recipes of one sub-group measure the difference between the total amounts of cations in the system and the amount of cations in the interstitial water; the former being obtained by, say, a barium or ammonium leach of a sample of wet or dry clay, and the latter from the product of the interstitial water volume and the concentrations of cations therein. However, equation (2) shows that this approach actually yields the sum of the cation surface-excesses. In freshwater clays the anion surface-excess will be much less than the CEC and, so the approach would offer a reasonable numerical approximation for CEC. However, the CEC will be under-estimated if the anion surface-excess is not negligible as is the case, for example, in saline or estuarine clays (Thomas et al., 1983). Since anion surface-excess can be as high as 200 meq kg^{-1} dry clay in sea water, an error in CEC as large as 20% can be involved; this has not been generally appreciated (Murthy and Ferrell, 1972; Devine et al., 1973).

Recipes of the other sub-group measure the decrease in concentration of a leachate cation, eg. barium or ammonium, observed when the leachate is mixed with a sample of clay (wet or dry). The apparent loss of leachate cation from solution is taken to be the amount of cation lost to the clay. The CEC has been calculated as the difference between the amount of leachate-cation present initially in the leachate, and that present in the final solution, calculated from the product of the volume and leachate-cation concentration of the bulk solution. Since all the leachate cation in the final mixture arises from the original leachate, the preceding calculation actually yields the surface excess of the leachate cation (equation 2). Moreover, since the leachate cation is deliberately present at high concentrations, this value will closely approximate the sum of the cation surface-excesses of the mixture. Consequently, the value of the CEC measured by this method would be under-estimated by an amount equal to the sum of the surface excesses of the anions in the mixture; as the leachate anion is present at high concentration its surface excess will closely approximate this value. As explained this can invoke errors in CEC as large as 20%. In contrast to the recipes of the previous sub-group, which only incur serious error with saline clays, this sub-group's recipes will under-estimate CEC of a given clay to very nearly the same extent irrespective of salinity. This is because the overwhelming effect of a high leachate - salt concentration will tend to maintain the same leachate - anion surface-excess.

ACKNOWLEDGEMENTS

This work was undertaken as part of the Natural Environment Research Council's "Geochemical Cycling" project.

REFERENCES

Avery, B.W. and Bascomb, C.L., 1974. "Soil Survey Tech. Mono." No.6, Harpenden.
Babcock, K.L. 1963. Theory of the chemical properties of soil colloidal systems at equilibrium, Hilgardia 34:417.

Bard, Y., 1971. "Non-linear Parameter Estimation", Academic Press, New York.
Berner, R. A., 1971. "Principles of Chemical Sedimentology", McGraw-Hill, New York.
Van Bladel, R. and Laudelout, H., 1967. Apparent irreversibility of ion exchange reactions in clay suspensions, Soil Sci. 104: 134.
Bolt, G. H., 1955. Analysis of the validity of the Gouy-Chapman theory of the electric double layer, J. Colloid Sci. 10:206.
Bolt, G. H., 1967. Cation-exchange equations used in soil science - A review, Neth. J. Agric. Sci. 15:81.
Bolt, G. H. and Bruggenwert, G. M., 1976. "Soil Chemistry. A. Basic Elements", Elsevier, Amsterdam.
Bower, C. A., Reitemeier, R. F. and Firman, M., 1952. Exchangeable cation analysis of saline and alkaline soils, Soil Sci. 73: 251.
Chapman, D. L., 1913. A contribution to the theory of electro-capillarity, Phil. Mag. 25:475.
Chapman, H. D., 1965. Cation exchange capacity, in: "Methods of Soil Analysis" Black, C. A., Evans, D. D., Ensminger, L. L., White, J. L. and Clark, F. E., eds., Amer. Soc. Agronomy, USA.
Davis, L. E., 1942. Significance of Donnan equilibrium for soil colloid systems, Soil Sci. 54:199.
Davis, L. E., 1945. Theories of base-exchange equilibriums, Soil Sci. 59;379.
Devine, S. B., Ferrell, R. E. and Billings, G. K., 1973. The significance of ion exchange to interstitial solutions in clayey sediments, Chemical Geology 12:219.
Dewis, J. and Freitas, F.,1970. "Physical and Chemical Methods of Soil and Water Analysis", FAO, Rome.
"Dictionary of Science and Technology", 1974, Chambers, UK.
Donnan, F. G., 1935. Molar (micellar) mass, electrovalency of ions and osmotic pressure of colloidal electrolytes, Trans. Faraday Soc. 31:80 (1935).
Gast, R. G., 1977. Surface and colloid chemistry, in: "Minerals in Soil Environments", Dixon, J. B. and Weed, S. B. eds., Soil Sci. Amer., Wisconsin.
Glasstone, S.,1962. "Textbook of Physical Chemistry", 2nd edition. Macmillan, London.
Gouy, G.,1910. Sur la constitution de la charge electrique à la surface d'un electrolyte, J. Physique 9: 457.
Hatch, F. H. and Rastall, R. H., revised by Greensmith, J. J.,1965. "The Petrology of the Sedimentary Rocks". Allen, London.
Helmholtz, O., 1879. Studien über Electrische Grenzschichten, Annal. Physik under Chemie 7: 337.
Helmy, A. K.,1963. On cation-exchange stoichiometry, Soil Sci. 95: 204.
Hesse, P. R.,1971. "A Textbook of Soil Chemical Analysis", Murray, London.

Isaac, R. A. and Kerber, J. D., 1971. Atomic absorption and flame photometry; Techniques and uses in soil, plant and water analysis, in "Instrumental Methods for Analysis of Soils and Plant Tissue", I. M. Walsh, ed., Soil Sci. Soc. Amer., Wisconsin.

Kelley, W. P., 1948. "Cation Exchange in Soils", Reinhold, New York.

Kruyt, H. R., 1952. "Colloid Science" Vol.1, Elsevier, Amsterdam.

Laudelout, H., 1965. A unified treatment of two ion exchange formulations commonly used in soil science, in: "Technical Reports Series", No. 48:20-24, International Atomic Energy Agency, Vienna.

Laudelout, H., Van Bladel, R., Gilbert, M. and Cremers, A., 1968. Physical chemistry of cation exchange in clays, 9th Int. Congr. Soil Sci. Trans. 565.

Laudelout, H., Van Bladel, R., Bolt, G. H. and Page, A. L., 1968. Thermodynamics of heterovalent cation exchange reactions in a montmorillonite clay, Trans. Faraday Soc. 64: -477.

Marshall, C. E., 1964. "The Physical Chemistry and Mineralogy of Soils", Wiley, New York.

Moore, W. J., 1974. "Physical Chemistry", Longmans, USA.

Murthy, A. S. P. and Ferrell, R. E. 1972. Comparative chemical composition of sediment intersititial waters, Clays and Clay Minerals 20:317.

Neal, C. 1977. The determination of adsorbed Na, K, Mg and Ca on sediments containing $CaCO_3$ and $MgCO_3$, Clays and Clay minerals 25:253.

Neal, C., Thomas, A. G. and Truesdale, V. W., 1982. The thermodynamic characterisation of clay electrolyte systems, Clays and Clay Minerals 30:291.

van Olphen, H., 1977. "An Introduction to Clay Colloid Chemistry", 2nd edition, Wiley, New York.

Overbeek, J. Th. G., 1956. The Donnan Equilibrium, Prog. Biophys. 6:58.

Overbeek, J. Th. G., and Lijklema, J., 1959., Electric potentials in colloid systems, in: "Electrophoresis", M. Bier, ed. Academic Press, New York.

Posner, A. M. and Quirk, J. P., 1964. The adsorption of water from concentrated electrolyte solutions by montmorillonite and illite, Proc. Roy. Soc. 278:35.

Reitemeier, R. F., 1946. Effect of moisture content on the dissolved and exchangeable ions of soils of arid regions, Soil Sci. 61:195.

Russell, K. L., 1970. Geochemistry and halmyrolysis of clay minerals, Rio Ameca, Mexico, Geochim. Cosmochim. Acta 34:893

Schofield, R. K. and Talibuddin, O., 1948. Measurement of internal surface by negative adsorption, Discussion Faraday Soc. 3:51.

Thomas, H. C., 1965. Toward a connection between ionic equilibrium and ionic migration in clay gels, in: "Technical Reports Series", No. 48: 4-19, International Energy Agency, Vienna.

Thomas, A. G., Truesdale, V. W. and Neal, C. The heterogeneous distribution of anions and water around a clay surface with special reference to estuarine systems. This volume, page 17.

US Dept. Agric., 1969. "Agricultural Handbook No. 60".

White, R. E., 1979. "Introduction to the Principles and Practice of Soil Science", Blackwells, Oxford.

Wiklander, L., 1964. Cation and anion exchange phenomena, in: "Chemistry of the Soil", F. T. Bear, ed., Reinhold, Holland.

Willard, H. H., Merritt, L. L. and Dean, D. A., 1965. "Instrumental Methods of Analysis", Van Nostrand, USA.

Yong, R. N. and Warkentin, B. P., 1966. "Introduction to Soil Behaviour" Macmillan, New York.

Zaytseva, E. D., 1962. Exchangeable cations in sediments of the Black Sea, Tr. Inst. Okeanol. 54:48.

THE HETEROGENEOUS DISTRIBUTION OF ANIONS AND WATER AROUND A CLAY SURFACE WITH SPECIAL REFERENCE TO ESTUARINE SYSTEMS

A. G. Thomas, V. W. Truesdale and C. Neal

Institute of Hydrology, Wallingford, Oxon, UK

INTRODUCTION

This paper presents new information on the distribution of anions and water around clays in the estuarine environment. This information is not already available for two main reasons. Firstly, previous studies of this kind were confined to systems containing homoionic clays and simple electrolytes, and the information thereby obtained is consequently of limited application to the natural environment. Secondly, other more general clay/electrolyte studies, which might have been expected to be able to supply the required information, cannot because they failed to consider the possibility of a non-uniform distribution of species in solution.

In this study a new variable, notional interfacial content (NIC) is used to express the distribution of species around the clay. Here, this is determined after measuring negative adsorption, that is, the increase in concentration of anions in solution that occurs when a dry, salt-free clay is added to an electrolyte solution. The advantage of using NIC over earlier approaches, eg. ones using ion surface-excess or even empirical ones, is that it accommodates any distribution, uniform or non-uniform, for the species, and that it is rigorously defined.

The use of NICs has shown that in estimating cation exchange capacity (CEC) and pore-water composition in clay/electrolyte systems, the effect of the non-uniform distribution of anions and water around the clay must not be ignored, as has been commonly the case in the past. Estimates of the error accruing from this in earlier work are presented.

THEORY

Only a brief description of the theory used in this paper will be given here. A complete description may be found elsewhere (Neal et al., 1982; Truesdale et al., 1983). The same nomenclature is adopted here to avoid confusion.

Cation Exchange Capacity

The CEC of a clay is defined as the amount of cations equivalent to the negative charge upon unit mass of the clay and is given by

$$CEC = T_+ - T_- \tag{1}$$

where T_+ is the total amount of cations and T_- the total amount of anions in the system.

Notional Interfacial Content

The notional interfacial content of any species i with respect to a chosen reference species j, $NIC_{i(j)}$, is given by

$$NIC_{i(j)} = T_i - T_j\, C_{i(j)} \tag{2}$$

where T is the total amount of i or j in the system, and $C_{i(j)}$ is the concentration of i with respect to j in the bulk solution.

The NIC concept can be readily related to negative adsorption. Thus application of equation (2) to the equilibrium mixture of clay plus electrolyte solution yields:

$$NIC_{i(j)f} = T_i - T_j\, C_{i(j)f}$$

where the subscript, f, refers to the final (ie. clay and electrolyte solution) mixture. Since it is assumed that neither anions nor water were introduced along with the clay,

$$T_i = T_j \cdot C_{i(j)o}$$

where o refers to the original electrolyte solution, ie. before addition of the clay. Combining these last two equations

$$NIC_{i(j)} = T_j\, (C_{i(j)o} - C_{i(j)f})$$
$$= -T_j\, (C_{i(j)f} - C_{i(j)o}) \tag{3}$$

where $(C_{i(j)f} - C_{i(j)o})$ is the negative adsorption of i.

It is worth pointing out here that NICs tell us only the distribution of one species <u>relative</u> to another. Thus Cl[−] NIC(W) gives the distribution of chloride relative to water; it does not provide a real or absolute spatial distribution.

METHODOLOGY

Experiments in which dry, salt-free clays (approximately 0.5 g) were mixed with various estuarine waters (5.00 ml) for sixteen hours were used to determine chloride NIC (water) and water NIC (chloride). The former was determined by means of equation (3) after measurement of the total water content of the system (equal to that present in the original electrolyte) and the concentration of chloride (with respect to water) both before and after addition of the clay. The latter was determined using the same measurements but the following equation,

$$NIC_W(Cl^-) = T_W \left(\frac{C_{Cl^-(W),f} - C_{Cl^-(W),o}}{C_{Cl^-(W),f}} \right) \quad (4)$$

which can be derived from equation (3) using the identities,

$$T_{Cl^-} = T_W \cdot C_{Cl^-(W),o}$$

and

$$C_i(j) = \frac{1}{C_j(i)}$$

Two sets of experiments were performed. One to establish the variation of the two NICs with changing salinity and two purified clays, another to establish the magnitude of these variables for a set of marine clays in artificial sea water (a.s.w). The first set of experiments used pure, homoionic (sodium) forms of an English kaolinite (Supreme China Clay) and of a smectite (Wyoming bentonite) both supplied by BDH Chemicals Ltd (UK). These clays were treated so as to obtain samples less than two microns mean spherical diameter, free of salts and acid soluble contaminants, and in a homoionic form (Posner and Quirk, 1964). However, to avoid chloride contamination, sodium acetate was used as the displacing agent. The second set used a variety of API standard smectites and illites. Kaolinites were not included because the NICs obtained for them in the first set of experiments were not significantly different from zero. For the sake of uniformity the API standard clays were converted into marine analogues by frequent washing with a.s.w. (Neal, 1977). As it was not possible to wash these clays and retain them in their marine form (Sayles and Manglesdorf, 1977), an estimate of chloride contamination had to be made. This chloride blank was performed by substituting distilled water for sea-water. The artificial estuarine water and a.s.w. were made from the recipe of Lyman and Fleming (1940), suitably diluted.

The CECs of all the clays were determined using a multiple-salt-extraction technique based on an earlier method (Neal, 1977) which eliminates the contributions from soluble and sparingly soluble salts. In effect this determines T_+ and T_- of equation (1).

Chloride was analysed using an automated version (Cook and Miles, 1979) of an earlier colorimetric method (Zall et al., 1956). Since small differences in large concentrations were often being sought, the chloride analyses were carried out in quintuplicate so as to reduce analytical error. The concentration of water (gl^{-1}) in the bulk solution was assumed not to change with the addition of the clay and was interpolated from data (Riley and Chester, 1971) for the variation of the specific gravity anomaly with salinity. This assumption introduces an insignificant error. Changes in sulphate concentration were undetectable with the automated version (Cook and Miles, 1979) of a method which used thorin and barium chloranilate (Persson, 1975). This was anticipated from earlier results (Oldham, 1975) and the fact that sulphate represents only approximately 10% of the total anion load (in equivalents l^{-1}) of estuarine- and sea-water. As all other anions even taken collectively, are of less significance than sulphate, chloride can be assumed to represent the total anion load. Thus, chloride NIC (water) is equivalent to anion NIC (water), the variable which, strictly, ought to have been determined.

There is no absolute definition of a dry clay and a wide variety of drying temperatures has been used, eg. 350^0C for 16 hrs (Posner and Quirk, 1964), 100^0C over a water bath (Helmy et al., 1980), air temperature (Bower and Goertzen, 1955). Our dry clays were dried by heating them to constant weight at 60^0C. This is a commonly used procedure which has the advantage of avoiding the more extreme conditions.

Some workers (eg. Bolt and Warkentin (1958), Edwards and Quirk (1962), de Haan and Bolt (1963)) prefer to start their experiments with the sample of clay already in contact with the electrolyte solution. In this way any irreversibility caused by heating to dryness and loss of exchange capacity from washing, may be avoided. However, others (Posner and Quirk, 1964; Bower and Goertzen, 1955; Schofield and Talibuddin, 1948) use dry clays and there appears to be no significant difference in the magnitude of the results from either approach.

The precision of these experiments is likely to be low since, as explained above, small differences in large values were often being sought. Consideration of the analytical errors involved suggests that the given values of negative adsorption are probably correct to $\pm$ 25 meq kg^{-1}.

To ensure that the practical procedure used here (preparation of clays; drying temperature) was compatible with ones used earlier, some of the earlier experiments with simple electrolytes were repeated. The data thus obtained were not only of a comparable magnitude but also, by application of the Schofield equation (Schofield, 1947) (see later) gave a similar estimate (5.4×10^5 m^2 kg^{-1}) for the surface area of a smectite.

RESULTS

Tables 1 and 2 present the chloride NIC (water) and water NIC (chloride) data for the estuarine and sea water studies. For the estuarine study with the pure, homoionic smectite the values of both chloride NIC (water) and water NIC (chloride) show a general decrease with increasing ionic strength, from -40 to more than 15,000 meq kg^{-1}, respectively, at a salinity of 5°/oo to -100 to less than 5000 meq kg^{-1}, respectively, at higher salinities. Values for the corresponding kaolinite were all zero within experimental error. The results for the standard clays in a.s.w. show a range of values from 0 to -223 meq kg^{-1} for chloride NIC (water) and from 29,000 to 0 meq kg^{-1} for water NIC (chloride).

Given that the NIC concept purports to differentiate between species "belonging" to the clay and the solution, some readers

Table 1. Values of Chloride NIC (Water)

	Standard Clays		Estuarine Clays	
Type	Cl$^-$ NIC(W) (meq kg^{-1})	Chlorinity (°/oo)	Cl$^-$ NIC(W) (meq kg^{-1})	
AP1 22B	- 90	2.7	- 40	
AP1 25	-140	5.4	- 55	
AP1 23	-110	8.1	-100	
AP1 21	-150	10.8	-110	
AP1 27	- 7	13.2	-140	
AP1 36	- 24	16.0	-120	
AP1 37	- 0	19.0	- 80	
AP1 24	-220			
AP1 22A	-180			
AP1 42	-120			
AP1 31	- 22			
AP1 41	0			
Bentonite	- 28			

Table 2. Values of Water NIC (Chloride)

Type	Standard Clays		Estuarine Clays	
	W NIC(Cl$^-$) (meq kg^{-1})	Chlorinity ($^o/oo$)	W NIC(Cl$^-$) (meq kg^{-1})	
APl 22B	11,000	2.7	15,000	
APl 25	16,000	5.4	10,000	
APl 23	14,000	8.1	13,000	
APl 21	17,000	10.8	11,000	
APl 27	2,500	13.2	11,000	
APl 36	4,900	16.0	8,000	
APl 37	0	19.0	4,500	
APl 24	29,000			
APl 22A	20,000			
APl 42	14,000			
APl 31	4,500			
APl 41	0			
Bentonite	5,700			

might suspect the trends in data and the two NICs. Thus, firstly, it might appear that increasing the chloride concentration should increase the amount of chloride "belonging" to the clay, and hence increase the chloride NIC (water). Secondly, the similarity in the trends in both sets of data might appear incompatible if an inverse relationship between the NICs is assumed. However, such an analysis overlooks the important fact that NICs offer a relative, not absolute, distribution; a decreasing chloride NIC (water) (notional system) is not incompatible with more chloride being nearer to the surface in the real system. Moreover, the similarity in the trends is explicable by the identity,

$$NIC_i(j) = -c_i(j) \cdot NIC_j(i) \tag{6}$$

which is derived using equations (2) and (5). Therefore, the two NICs are not inversely related. Moreover, since the concentration of chloride (with respect to water) is always positive and increases almost linearly with the salinity of the bulk solution, the two types of NIC should not be expected to change in the same way with respect to salinity. The particular trends obtained for the two NICs are consistent with the concentration of chloride (with respect to water), $c_i(j)$, decreasing more rapidly than does chloride NIC (water) with decrease in salinity; water NIC (chloride) tends to infinity as the salinity tends to zero (equation (6)).

DISCUSSION

Errors

The above results can be used to calculate the errors which result from assuming that there is a uniform distribution of anions and water around the clay. It is explained elsewhere (Truesdale et al., 1983) how this tacit use of the Helmholtz model has come to pervade work upon clay/electrolyte interactions. Since chloride NIC (water) and water (NIC) chloride quantify the relative inhomogeneity in anion and water distributions in the real system, an assumption of an homogeneous distribution in both chloride and water (Helmholtz model) is a dismissal of the two NICs. Therefore the error accruing from the use of the Helmholtz model is directly proportional to the magnitude of the particular NIC that has been neglected.

By estimating the CEC of a clay under the assumption of a homogeneous distribution of species in solution many workers have, in effect, neglected chloride NIC (water). Since the CEC is given by,

$$CEC = \sum_{+} NIC_k(j) - \sum_{-} NIC_k(j) \qquad (7)$$

where k is an ion, it follows that neglect of chloride NIC (water) incurs a fractional error,

$$\varepsilon = \frac{NIC_{Cl^-}(W)}{CEC} \times 100\%$$

Values of ε for the estuarine smectite experiment (Table 3) show that the error is as large as 14% at a salinity of 33°/oo and is significant even at low salinities. Also, for five of the thirteen standard clays in a.s.w. the error was greater than 10%.

By estimating the total amount of an anion present in the system using the anion's concentration in the real bulk solution together with the assumption that the Helmholtz model is valid, many workers have, in effect, again neglected anion NIC (water). Thus, for the Helmholtz model the solution forms a single phase, the bulk solution, and consequently requires that anion NIC (water) is zero. Thus, the estimated total will be in error by an amount equal to anion NIC (water). This emphasises that the total content of each anion in the system must be determined experimentally, not calculated; measurement of anion NIC (water) requires that the total anion content be determined experimentally. This point would be important when calculating the anion content of the interstitial pore-water of a clay sediment. The discrepancy expressed as,

Table 3. Values of ε and E

Standard Clays				Estuarine Clays			
Types	ε (%)	E(%) 1:1 clay:solution	E(%) 1:10 clay:solution	Salinities (°/oo)	ε (%)	E(%) 1:1 clay:solution	E(%) 1:10 clay:solution
APl 22B	9.1	15	1.5	4.8	4	48	4.8
APl 25	13.8	23	2.3	9.5	5.5	33	3.3
APl 23	9.7	18	1.8	14.3	9.8	39	3.9
APl 21	12.1	25	2.5	19.0	11.2	33	3.3
APl 27	<1	1	<1	23.3	13.8	34	3.4
APl 36	7	4	<1	28.2	12.1	24	2.4
APl 37	0	0	0	33.5	8.2	14	1.4
APl 24	18.6	38	3.8				
APl 22A	18.3	31	3.1				
APl 42	25.2	21	2.1				
APl 31	1.5	4	<1				
APl 41	0	0	0				
Bentonite	2.3	5	<1				

$$E = \frac{\text{anion NIC (water)}}{T_{\text{anion}}} \times 100\%$$

has been evaluated from the experiments described above. The results (Table 3) show that for a 1:1 mixture by weight of dry smectite to estuarine water, E is $\simeq$ 50% at low salinities and between 14 and 40% at intermediate to sea water salinities. Further, for the 13 smectites exposed to sea water, in 5 cases E was greater than 20%. Note that, as shown in Table 3, if the sediment to interstitial water ratio had been much less (eg. 1:10), then E would be much smaller (eg. < 2% at sea water salinities) and hence under such conditions unimportant.

The method of estimating the effect of omitting water NIC (chloride) follows a parallel course to that involved above with the omission of chloride NIC (water). However, the omission of water NIC (chloride) is found to incur no error in the estimation of CEC because water, being an uncharged species, is not represented in equation (7). Nevertheless, in the same way that the omission of chloride NIC (water) incurs an error in estimates of the total chloride in the system when the Helmholtz model is invoked, omission of water NIC (chloride) would introduce an error into the estimate of the total amount of water in the system. Even so, this error is unlikely to be a problem because the total water content of the system will usually be measured, not calculated.

The Schofield Equation

The Schofield equation (Schofield, 1947) has been used successfully on several occasions to model the change in chloride surface-excess with change in electrolyte concentration (Bolt and Warkentin, 1958; Edwards and Quirk, 1962; de Haan, 1964). It seemed worthwhile, therefore, to attempt to adapt it to the estuarine NIC data. This presented two problems because the equation is limited to systems which have a constant water concentration throughout, and which contain only a single electrolyte. The former problem proved to be of little practical significance when it was found that the values of chloride surface excess and chloride NIC (water) for a kaolinite and a smectite (the same clays used in the earlier estuarine study), in a range of sodium chloride solutions (0-1.0 mol. l^{-1}) differed by less than 3%. It seems, therefore, that with these clays in this range of salt concentrations, the assumption of a constant water concentration throughout the system is valid. The latter problem was overcome by representing each sample of estuarine water by a single-electrolyte analogue whose molarity was equal to the total molarity of the estuarine sample.

The Schofield curves for 1:1, 1:2, 2:2 and 2:1 electrolytes are superimposed upon the plot of chloride NIC (water) versus salinity in Fig. 1. The closest fit between experimental and

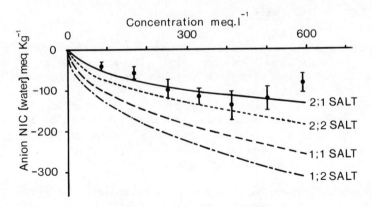

Fig. 1. Anion NIC (water) versus concentration for the estuarine set of experiments. The curves shown are for anion surface-excesses calculated from the Schofield equation and are for 2:1, 2:2, 1:1 and 1:2 salts. These salts are taken to be in equilibrium with a clay having a CEC of 1000 meq kg^{-1} and a surface area of 5.4×10^5 m^2 kg^{-1}.

Schofield-predicted data is obtained with the 2:1 electrolyte analogue. This is perhaps surprising given that 1:1 electrolytes dominate the composition of the artificial estuarine waters. One possible explanation is offered by the Gouy-Chapman model when it predicts that divalent ions are very effective at compressing the ion swarm nearer to the clay surface (Schofield, 1947). This compression would lower the values of chloride NIC (water).

Positive Sites

Strictly, it is as important to know the number of positive sites upon a clay as the number of negative sites since both are fundamental properties of the clay. However, historically, the latter has received more attention because it is numerically more important; "clays are cation exchangers". While an estimate of the number of positive sites upon a clay is difficult to obtain, a maximum value can be calculated from the above results. The approach used to determine NICs here assumes that the clay which is added to the electrolyte solution does not possess any positive sites. The effect of this assumption being invalid can be ascertained by including an appropriate term for the anions associated with the positive sites during derivation of equation (3). For example, in the case of chloride NIC (water) the total chloride content used in equation (2) will have to include an amount of chloride extra to that of the initial electrolyte solution. This process yields

$$NIC_{Cl^-(W)} = -T_W(C_{Cl^-(W),f} - C_{Cl^-(W),o}) + \lambda_{Cl^-} \qquad (8)$$

where λ_{Cl^-} is the number of positive sites occupied by chloride ions. Given that λ_{Cl^-} makes our measured values of chloride NIC (water) (Bolt and Warkentin, 1958) more negative than they should be, and that it is reasonable to assume that the correct chloride NIC (water) will always be less than or equal to zero, the value of λ_{Cl^-} cannot be greater than the magnitude of the modulus of chloride NIC (water). Moreover, since the number of positive sites, and hence λ_{Cl^-}, is a property of the clay not dependent upon salinity the modulus of the smallest value of chloride NIC (water) is the limit for λ_{Cl^-}. By this argument, then, the maximum number of positive sites on the smectite used in the estuarine experiment described above was 40 meq kg^{-1}. This figure can be compared with the CEC of approximately 1000 meq kg^{-1} and also with values of approximately 16 meq kg^{-1} (Bolt and Warkentin, 1958) and 20 to 60 meq kg^{-1} (Bolt and de Haan, 1965) for the number of positive sites on two similar clays, obtained using an analogous approach but with chloride surface excess.

The derivation of equation (8) may be extended to include the errors in $NIC_{i(j)}$ due to any incorrectness in the assumption that the clay which is added to the electrolyte is dry and totally free of anions. For example, if the clay were contaminated with sodium chloride the term λ_{Cl^-} of equation (8) would have to be expanded to include the further amount of chloride. This reasoning yields,

$$NIC_{i(j)} = -(T_j + \lambda_j)(C_{i(j),f} - C_{i(j),o}) + \lambda_i \qquad (9)$$

where λ_i and λ_j are the amounts of i and j causing errors in T_i and T_j respectively. This general equation does not differentiate between anions arising from contaminating salts or positive sites. The use of equation (8) to calculate the number of positive sites was only possible because the clay had been very thoroughly washed to remove salts. Equation (9) shows that the effects of λ_i and λ_j upon $NIC_{i(j)}$ are antipathetic.

Natural Sediments

So far the discussion has been limited to the problem of rigorously modelling clay/electrolyte systems in the laboratory. It would have been desirable, for the purposes of comparison, to have had a set of data from the natural environment which had been collected with the same rigour. Unfortunately such data do not exist and the set of data which comes closest to fulfilling this role is that obtained by Murthy and Ferrell (1973). These workers studied the effect of adding a natural marine sediment to various amounts of distilled water. While they claimed to be measuring "exchangeable cations" we show elsewhere (Truesdale et al., 1983) that the method they used actually measured something close to the cation surface-excesses. Most interestingly, values of chloride NIC (water) inferred from Murthy and Ferrell's results increase with increasing salinity, that is, they present the opposite trend to both that encountered in our laboratory studies and that for anion surface excesses predicted by the Schofield equation. Unfortunately, this striking difference between the behaviour of the laboratory clays and natural sediments is open to question. This is because the inference of chloride NIC (water) values from the original data requires the assumptions that the CEC of the sediment was constant (as Murthy and Ferrell assumed) and that the values of cation surface-excess approximate those of cation NIC (water). It is also possible that organic coatings or the flocculation of sediment particles might be responsible for this difference in trend.

ACKNOWLEDGEMENTS

This work was undertaken as part of the Natural Environment Research Council's "Geochemical Cycling" project.

REFERENCES

Bolt, G. H. and de Haan, F. A., 1965, Interactions between anions and soil constituents, in: "Technical Reports Series", 48:94, International Atomic Energy Agency, Vienna.

Bolt, G. H. and Warkentin, B. P., 1958, The negative adsorption of anions by clay suspensions, Kolloid Z., 156:41.

Bower, C. A. and Goertzen, J. O., 1955, Negative adsorption of salts by soils, Soil Sci. Amer. Proc. 19:147.

Cook, J. M. and Miles, D. L., 1979, Methods for the analysis of groundwater, Institute of Geological Sciences, Internal Report No. WD/ST/79/5.

Edwards, D. G. and Quirk, J. P., 1962, Repulsion of chloride by montmorillonite, J. Colloid. Sci., 17:872.

de Haan, F. A. M., 1964, The negative adsorption of anions (anion exclusion) in systems with interacting double layers, J. Phys. Chem., 68:2970.

de Haan, F. A. M. and Bolt, G. H., 1963, Determination of anion adsorption by clays, Proc. Soil Sci. Soc. Amer., 27:636.

Helmy, A. K., Natale, I. M. and Manolesi, M. E., 1980, Negative adsorption in clay-water systems with interacting double layers, Clays and Clay Minerals, 28:262.

Lyman, J. and Fleming, R. H., 1940. Composition of sea water, J. Mar. Res., 3:134.

Murthy, A. S. P. and Ferrell, R. E., 1973, Distribution of major cations in estuarine sediments, Clays and Clay Minerals, 20:317.

Neal, C., 1977, The determination of adsorbed Na, K, Mg and Ca on sediments containing $CaCO_3$ and $MgCO_3$, Clays and Clay Minerals, 25:253.

Neal, C., Thomas, A. G. and Truesdale, V. W., 1982, The thermodynamic characterisation of clay electrolyte systems, Clays and Clay Minerals, 30:291.

Oldham, K. B., 1975, Composition of the diffuse double layer in sea water or other media containing ionic species of +2, +1, -1 and -2 charge types, J. Electroanal. Chem., 63:139.

Persson, G. A., 1975, Automatic colorimetric determination of low concentrations of sulphate for measuring SO_2 in ambient air, Air and Water Pollut. Int. J., 10:139.

Posner, A. M. and Quirk, J. P., 1964, The adsorption of water from concentrated electrolyte solutions by montmorillonite and illite, Proc. Roy. Soc., 278:35.

Riley, J. P. and Chester, R., 1971, "Introduction to Marine Chemistry", Academic Press, London.

Sayles, F. L. and Manglesdorf, P. C., 1977, The equilibrium of clay minerals with sea water: exchange reactions, Geochim. Cosmochim. Acta, 41:951.

Schofield, R. K., 1947, Calculations of surface areas from measurements of negative adsorption, Nature, 160:408.

Schofield, R. K. and Talibuddin, O., 1948, Measurement of internal surface by negative adsorption, Faraday Soc. Discus., 3:51.

Truesdale, V. W., Neal, C. and Thomas, A. G., 1983, A new perspective of several approaches to clay electrolyte studies. This volume, page 1.

Zall, D. M., Fischer, D. and Garner, M. Q., 1956, Photometric determination of chlorides in water, Anal. Chem., 28:1655.

HUMIC SUBSTANCES AND THE SURFACE PROPERTIES OF IRON OXIDES IN FRESHWATERS

E. Tipping

Freshwater Biological Association, The Ferry House
Ambleside, Cumbria, UK

INTRODUCTION

Adsorption by iron oxides in natural waters, sediments and soils is commonly studied in order to understand the biogeochemical behaviour of the adsorbing species, or adsorbate. Prominent examples of such adsorbates are inorganic phosphate (see e.g. Hingston et al. 1968; Ryden et al. 1977) silicate (Hingston et al. 1968; Mott 1970) and trace metals (see e.g. Jenne 1968; Gadde & Laitinen 1974; Forbes et al. 1976). Another reason for studying adsorption is to see how it affects the particulate material itself, by influencing the surface presented to the environment. It is this surface which decides rates and extents of physical association with other particles - and consequently settling rates, grain size etc. In addition, processes such as crystallization and dissolution, as well as catalytic properties, are surface dependent.

In principle, understanding of both aspects requires the same information; knowledge of all the adsorbates present, the amounts of each adsorbed, and the surface topography. However the nature of the surface presented to the environment may be determined by only some of the adsorbates, if those adsorbates have properties which allow them to dominate the others. In this paper evidence is presented to show that for iron oxides in freshwater environments such a role may be played by humic substances.

Humic substances are the refractory, condensed, decomposition products of living matter, chiefly plants and algae. They consist mainly of C, H, N and O. Their formation, distribution, composition and properties have been reviewed by, among others, Schnitzer & Khan

(1972), Jackson (1975) and Gjessing (1976). Table 1 summarizes some properties of humic substances isolated from three chemically and biologically different British lakes. With regard to the work described in this paper the important properties are the dissociating groups - probably carboxyl (pK $\sim$ 5) and phenolic (pK $\sim$ 10) - and the fairly high molecular weights.

ADSORPTION STUDIES

Simple Electrolyte Media (NaCl)

Measurements have been made of the adsorption of Esthwaite Water humics by several model iron oxides (Tipping 1981a) and of the adsorption of humics from Esthwaite Water, Penwhirn Reservoir and Rostherne Mere to a single sample of goethite, α-FeOOH (Tipping 1981b). In all cases the adsorption isotherms approximate fairly closely a Langmuir isotherm when dilute NaCl is the background electrolyte. This means that α, the amount of humics adsorbed, and c, the free humic concentration are related by the following equation:

$$\alpha = nKc/(1+Kc) \qquad (1)$$

where n is the adsorption capacity and K the apparent equilibrium constant. Plots of α against c at two different pH-values are shown in Fig. 1. Both n and K decrease with increasing pH. Isotherms for the adsorption by goethite of soil-derived humic and fulvic acids also approximate Langmuir isotherms (Parfitt et al., 1977).

A probable mode of interaction of the humics with the oxide surface is ligand exchange whereby humic anionic groups replace surface-coordinated OH^- and H_2O (Parfitt et al. 1977). Regarding the geometry of the adsorbed humic molecules, some insight was gained from comparative measurements of the adsorption to a single sample of goethite of humics from the three lakes of Table 1 (Tipping 1981b). It was found that although the mass adsorption capacity, n (mg humics/g oxide), increased with increasing molecular weight, the molar adsorption capacity, n_M (moles humics/g oxide), did not. Furthermore the values of n or n_M were too large to be consistent with the humics being adsorbed as close-packed spheres. There was no dependence on molecular weight of the apparent molar equilibrium constant for adsorption, K_M. These results can be rationalized by a model in which, irrespective of molecular weight, each type of humic molecule has essentially the same number of adsorptive contacts with the oxide surface. This would account for the approximately equal values of K_M. The fraction of a humic molecule not in contact with the oxide increases

TABLE 1. Aquatic humic substances from three British lakes
(from Tipping 1981b)

		Esthwaite Water	Penwhirn Reservoir	Rostherne Mere
Lake concn. (mg/litre)		1-3	30-45	3-4
Composition C (wt. %)		53.6	56.7	61.1
	H	6.4	5.3	6.3
	N	2.6	1.5	1.3
Extinction coeffs. $E^{1\%}_{1cm}$	250nm	202	154	236
	340nm	67	73	69
Molecular weights*	M_w	5200	13400	4800
	M_n	3100	6900	3300
Dissociating groups mequiv/g	pK ~ 5	5	4	7
	pK ~ 10	≥ 5	≥ 1	≥ 3
Adsorption to goethite†				
n (mg humics/g goethite)		30.1	59.0	16.8
n_M (μmol humics/g goethite)		5.8	4.4	3.5
K_M (litre/μmol humics)		1.6	3.5	3.0

* Estimated by gel filtration on Sephadex G75 eluted with 0.1M NaOH. M_w = wt.ave.m.wt. M_n = no.ave.m.wt.

† At pH 7, 0.01M NaCl.

with increasing molecular weight, thus accounting for the dependence of n - but lack of dependence of n_M - on molecular weight. In terms of this model the observed Langmuir behaviour of the adsorption isotherm suggests that the mode of interaction between the humics and the oxide is independent of α, i.e. that the number of contacts per humic molecule does not change - a decrease might be expected - on going from low to high adsorption density.

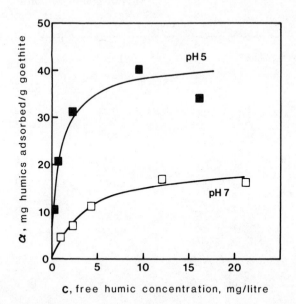

Fig. 1. Adsorption of Esthwaite Water humic substances by goethite at pH 5 and pH 7 in 0.002M NaCl. The curves are fitted to equation (1).

Effect of Bivalent Cations

In an electrolyte medium similar to the major-ion composition of Esthwaite Water (EW), humic adsorption by iron oxides exceeds that in NaCl. The difference becomes greater as more humics adsorb. The effect is illustrated in Fig. 2. Analysis of the adsorption isotherms (Tipping 1981a) shows that the enhanced adsorption results from a greater adsorption capacity. The apparent equilibrium "constant" for adsorption in the presence of EW major ions falls with increasing adsorption. The effect is due to the bivalent cations in the medium, Mg^{2+} and Ca^{2+}. Electrophoretic results (see below) suggest that as α increases, progressively more bivalent cations are coadsorbed.

A possible mechanism to account for these observations is that the bivalent cations compete with the oxide surface for humic functional groups (humics are known to be able to form complexes with Ca^{2+} and Mg^{2+} in solution, see e.g. Schnitzer & Hansen 1970). This could result in the humics making fewer contacts with the surface, allowing more molecules to adsorb (n increases) but with lower affinity (K decreases). Also in the presence of the cations humic adsorption may take place via oxide-cation-humic bridges as well as directly to the oxide.

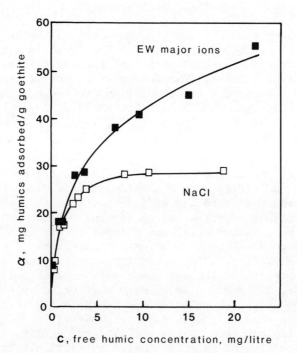

Fig. 2. The effect of electrolytes on the adsorption of Esthwaite Water humic substances by goethite, pH 7, I = 0.002M. the enhancement in the presence of Esthwaite Water major ions is due to Ca^{2+} and Mg^{2+}. Note that the results refer to a different sample of goethite to those in Fig. 1.

Competing Adsorbates

Anions which adsorb specifically to iron oxide, and which are present in natural waters, include phosphate, silicate and sulphate. The first two are able to compete with humics for adsorption (Table 2), although significant reduction of humic adsorption requires the competitors to be present at high concentrations. Sulphate has no detectable effect, even at concentrations as high as 0.03M (I = 0.7M, pH 7).

ELECTROPHORESIS STUDIES

The technique of electrophoresis relies on the fact that a charged particle in a liquid will move under the influence of an electric field (for reviews see Shaw 1969; Ottewill & Holloway 1975). The experimentally measured quantity is the electrophoretic mobility, u, i.e. the particle velocity divided by the voltage gradient. Mobilities are usually measured by direct observation of the particles by dark field microscopy.

Two approaches to the study of geochemical problems by electrophoresis may be distinguished. The first is simply to measure mobilities of natural particles (see e.g. Neihof & Loeb 1972, Hunter & Liss 1979). The second is to add well-characterized model particles to water samples and to see how their mobilities are affected (see e.g. Neihof & Loeb 1972, 1974; Loeb & Neihof 1975; Hunter 1980; Tipping 1981a; Tipping & Cooke 1982). The model particles can be regarded as probes for natural adsorbates.

In simple systems the plane of electrokinetic shear can reasonably be identified with the Stern surface, i.e. the boundary between the fixed and diffuse parts of the double layer. This allows theories for converting u to the shear potential to be combined with double layer theory in order to calculate charge densities. However for particles to which high molecular weight, flexible polymers are adsorbed the identification of the shear potential with the Stern potential is not justified (Lyklema 1978). It remains to be seen whether aquatic humic substances are sufficiently large and flexible for this restriction to apply to the results presented here, however for present purposes we simply assume that a positive or negative value of u indicates a positive or negative charge respectively at the shear plane, and that the greater is u the greater is the charge density. Thus the sign and magnitude of u gives a direct, but only qualitative, indication of the surface charge of the particles under study.

TABLE 2. Effects of phosphate and silicate on humic adsorption to goethite (from Tipping 1981a)

additions	humics adsorbed (mg/g goethite)	relative adsorption
none	23.5	100
10^{-6} M phosphate	21.2	90
10^{-5} M phosphate	11.2	47
10^{-4} M phosphate	5.0	21
5.4×10^{-6} M silicate	22.4	95
1.7×10^{-4} M silicate	19.2	82

<u>Total concentrations</u>: goethite 0.1 g/litre,
 EW humics 4 mg/litre

<u>Electrolyte medium</u>: EW major ions + NaCl, I = 0.002M.

Iron Oxides in Esthwaite Water (EW)

Fig. 3(a) shows the dependence of u on pH for a sample of synthetic goethite in NaCl. The change in u from large and positive at low pH to large and negative at high pH reflects the transition of surface $FeOH_2^+$ groups through FeOH to FeO^-.

In Fig. 3(b) the effect of the major ions of EW at their lakewater concentrations is shown. The general shape of the u vs. pH plot is the same as in Fig. 3(a), but the mobilities are smaller in magnitude, due to specific adsorption of SO_4^{2-} at low pH and of Ca^{2+} and Mg^{2+} at high pH.

Fig. 3(c) shows the effect of humic substances. In both NaCl and in the EW major ion medium a dramatic change in the u vs. pH plot is brought about by their adsorption to the goethite, the mobilities being negative over the whole pH range studied. The effect of the humics is not directly proportional to their concentration. Thus, as shown in Fig. 3(c), in the presence of EW major ions essentially the same u vs. pH plot is obtained at

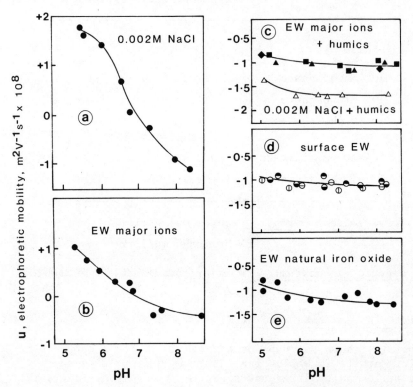

Fig. 3. Electrophoretic mobility vs. pH for synthetic iron oxides in various media (a-d) and for naturally-occurring iron oxide from Esthwaite Water (EW) (e).

The results in (a), (b) and (c) refer to goethite. In (c) the symbols refer to humic concentrations in mg/litre as follows: ■ 0.6, ◆ 1.75, ▲ △ 2.9. In (d) the results are for goethite added to surface water sampled in July 1979, humic concentration 1.6 mg/litre ◐ , and added to surface water sampled in November 1979, 2.1 mg/litre ◑ , and for haematite ⊕ and amorphous Fe-gel ⊖ added to water sampled in November 1979.

total humic concentrations in the range 0.6-2.9 mg/litre corresponding to a range of values of α of approx. 5 - 20 mg/g goethite at pH 7. The greatest changes in u are those brought about by humics adsorbing at low values of α. From experiments in which α and u were measured simultaneously (Tipping 1981a) the parameter $\Delta u/\Delta \alpha$ ($m^2 V^{-1} s^{-1}$/mg humics/g oxide) can be calculated. In 0.002M NaCl at pH 7, $\Delta u/\Delta \alpha$ = 0.6 at α ≃ 0, but only 0.04 at α ≃ n. In a medium consisting of EW major ions the values corresponding to α ≃ 0 and α ≃ n are 0.25 and < 0.01 respectively. The progressive decrease in the amount of negative charge conferred on the surface by adsorbing humics as more of them adsorb is discussed below, and is illustrated by the plots in Fig. 4.

Comparison of Fig. 3(d) with 3(c) shows that the medium consisting of EW major ions + humics imitates very well the lakewater itself regarding the electrophoretic mobility of the suspended goethite particles. Similarly close agreement between the synthetic and natural lakewaters was found for particles of haematite (α-Fe_2O_3) and amorphous Fe-gel added to EW (Tipping 1981a) and for goethite with three other lakewaters : Penwhirn Reservoir, Rostherne Mere and Wastwater (Tipping & Cooke 1982). This is strong evidence that humics are important in determining the surface characteristics of iron oxide particles in natural waters. Further support comes from experiments on naturally occurring particulate iron from EW. The latter material is formed in the lake (water column or sediment, depending on whether or not the lake is thermally stratified) by oxidation of soluble ferrous iron at or near the oxic/anoxic boundary (see Davison et al. 1981 for a more detailed description). The material - described by Tipping et al. (1981) - consists mostly of particles of amorphous iron oxide with mean diameters ≲ 0.5 μm. Approximately 35% by weight is Fe and 18% is carbon. Because of "contamination" by bacteria, algal debris etc. it is unlikely that all the carbon is in the Fe-containing particles. However the 4-7% of the total weight which was shown to be humic carbon is probably associated with the iron. Electrophoretic measurements on water samples in which the particulate material consisted almost entirely of the natural oxide gave a u vs. pH plot very similar to those for model iron oxides suspended in surface samples of EW (compare Figs 3(e) and 3(d)).

Effects of Ca^{2+} and Mg^{2+} on Electrophoretic Mobilities

As indicated by the plots in Fig. 3(c) bivalent cations - Ca^{2+} and Mg^{2+} - are important in determining the actual magnitude of the negative electrophoretic mobility caused by adsorbed humics. This is more fully demonstrated by the results in Fig. 4 which encompass wider ranges of humic and Ca^{2+} concentrations. Additional experiments (Tipping & Cooke 1982) showed that the changes in mobility with changes in [Ca^{2+}] are not simply due to changes in ionic strength. There is specific adsorption of the bivalent cation, probably via interaction with humic functional groups not involved in actual adsorption to the oxide. This topic has already been mentioned with regard to adsorption in the presence of bivalent cations.

Effects of Phosphate and Silicate on Electrophoretic Mobilities

Both phosphate and silicate can compete with humics for adsorption to iron oxides (Table 2). However their effects on the

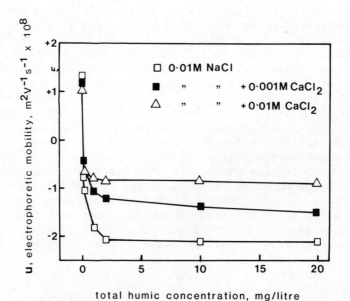

Fig. 4. Electrophoretic mobility of goethite: effects of humic and Ca^{2+} concentrations at pH 6.7. The concentration of goethite (10 mg/litre) was sufficiently low to make the total humic concentrations approximately equal to the free humic concentrations. The humics were from Penwhirn Reservoir.

electrophoretic mobilities of particles in the presence of humics are much less marked. This is illustrated by the results in Table 3. The lack of effect of silicate is perhaps not surpising since it is a poor competitor (Table 2). The effect of PO_4 is noticeable at 10^{-4}M in 0.01M NaCl, but not at all in 0.01M NaCl - 0.001M $CaCl_2$.

QUALITATIVE MODEL OF THE IRON OXIDE SURFACE WITH ADSORBED HUMICS

The experimental results summarized above allow a qualitative model of the iron oxide surface in the presence of humic substances to be put forward. From the results of adsorption experiments with humics of different molecular weights it is suggested that only a segment of a humic molecule actually interacts with the oxide surface (probably by ligand exchange with surface-coordinated OH^- or H_2O). Non-adsorbing parts of the molecule extend away from the surface, with their functional groups able to interact with other chemical species (Fig. 5). At all but the lowest values of α (see Fig. 4) it is these groups which determine the potential at the plane of electrokinetic shear, i.e. the potential 'sensed' by electrophoresis. A similar picture of the surfaces of solids to which seawater organic matter was adsorbed has been given by Hunter (1980). The ability of hydrophilic macromolecules to confer their own electrokinetic characteristics on particles to which they adsorb is well known (see e.g. Shaw 1969).

Several aspects of the electrophoresis results can be understood in terms of the humic groups not involved in adsorption to the oxide, and their interactions with protons and cations. Such interactions bring about decreases in the net negative charge density in the interfacial region and thus electrostatic repulsion betweem the anionic groups of adsorbed humics can be overcome. This explains the variation of $\Delta u/\Delta \alpha$ with α which is described above: humics adsorbing at low α can be positioned at the surface far enough apart not to repel each other electrostatically whereas those adsorbing at high α cannot and must take up either protons or cations to overcome the repulsion. (Uptake of protons or cations will be favoured in any case by the low dielectric strength in the interfacial region, and by the excess of positively charged species in the diffuse double layer due to the net negative charge of the fixed layer: cf. Rubio & Matijevic 1979; Davis & Leckie 1980.) The extent of the interactions of cations with humic functional groups in the shear plane obviously depends on the cation concentration, which explains the variation of u with $[Ca^{2+}]$ shown in Fig. 4. The uptake of Ca^{2+} and Mg^{2+} is an important factor in iron oxide-humic interactions since not only does it influence the fixed layer charge but also the extent of adsorption (see the ADSORPTION STUDIES section).

TABLE 3 Effects of phosphate and silicate on the electrophoretic mobility of goethite in the presence and absence of humic substances from Penwhirn Reservoir.

additions	electrophoretic mobility ($m^2V^{-1}s^{-1} \times 10^8$)	
	0.01M NaCl	0.01M NaCl + 0.001M $CaCl_2$
none	+ 1.24	+ 1.19
1 mg/litre humics	− 1.80	− 0.95
+ 2.5×10^{-6}M phosphate	− 1.72	− 1.02
+ 10^{-5}M phosphate	− 1.96	− 1.05
+ 10^{-4}M phosphate	− 2.03	− 0.93
+ 2.5×10^{-4}M silicate	− 1.76	− 0.91
10^{-4}M phosphate	− 2.24	− 0.67
2.5×10^{-4}M silicate	− 0.20	− 0.31

The concentrations shown are total concentrations, but because of the low concentration of goethite used (0.01g/litre) they can be taken as approximately equal to free (unadsorbed) concentrations. The pH was 6.7. The errors in the mobility values (s.e.m.) were ≤ $0.10 \times 10^{-8} m^2V^{-1}s^{-1}$.

The extent of ionization of humic functional groups in the shear plane allows the pH dependence of U to be accounted for (Fig. 3). The mobility either stays the same or becomes more negative as the pH increases (Fig. 4), despite the fact that at a fixed total humic concentration decreasing amounts of humic substances are adsorbed (Fig. 1). In the absence of bivalent cations the more negative mobility values are probably due to decreased protonation (i.e. increased ionization) of the functional

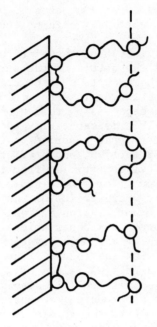

Fig. 5. Schematic picture of humic substances adsorbed at the iron oxide surface. Ionizable functional groups are denoted by ○. The dashed line is the electrokinetic shear plane.

groups in the shear plane, not only because of the fall in $[H^+]$ but also because the lower adsorption density lessens the need to overcome electrostatic repulsion between ionizable groups of adjacent adsorbed humic molecules. When Ca^{2+} and/or Mg^{2+} are present, the decreased protonation of the adsorbed humics due to decreases in $[H^+]$ can be countered by increased interaction of the humics with these cations. This buffering effect probably accounts for the near-independence of mobility on pH for the EW system (Fig. 3).

Similar arguments apply to the inability of phosphate and silicate to alter the electrophoretic mobility, despite their shared ability to compete with humics for adsorption (Tables 2 & 3). Since both of these adsorbates are of small molecular size neither will be in the shear plane when adsorbed, as long as a certain adsorption density of humics is maintained. Decreases in humic adsorption brought about by phosphate and silicate are countered by increased ionization of humic functional groups in the shear plane, due to diminished electrostatic repulsion. This would apply to any low molecular weight adsorbate. The model thus allows a complex speciation at the iron oxide surface in a natural water, but still provides a relatively simple description of the electrokinetic shear plane as long as some of the adsorbates are humic substances with relatively high molecular weights.

In the foregoing discussion the electrophoretic mobility has been taken to indicate the nature of the surface of the iron oxide particles, and while the results provide clear evidence for the importance of humic substances in determining the surface characteristics, it is worth emphasising that u only gives a general, qualitative picture. This is especially the case for a surface with flexible polyelectrolytes adsorbed to it, as mentioned in the ELECTROPHORESIS STUDIES section, and prediction of the colloid stability of the iron oxide particles from the electrophoretic results requires caution. For simple hydrophobic colloids the Derjaguin-Landau-Verwey-Overbeek (DLVO) theory (see Verwey & Overbeek 1948) offers a correlation between u and colloid stability. Indeed rudimentary observations of degrees of dispersion in the goethite-PR humics system of Fig. 4 were in qualitative agreement with the theory (Tipping & Cooke 1982). It was found that large (negative) values of u and low cation concentrations were associated with colloid stability, whereas small values of u and high cation concentrations resulted in aggregation. However deviations from DLVO theory are common in systems involving surfactants and/or polymers (see e.g. Gregory 1978) and therefore more rigorous determinations of colloid stability in relation to humic adsorption, cation concentrations and electrophoretic mobilities are needed.

ACKNOWLEDGEMENTS

The research described in this paper was funded by the Natural Environment Research Council. The final version was typed by Elisabeth M. Evans.

REFERENCES

Davis, J.A. and Leckie, J.O., 1980, Surface ionization and complexation at the oxide/water interface. 3. Adsorption of anions. J. Coll. Int. Sci., 74:32.
Davison, W., Heaney, S.I., Talling, J.F. and Rigg, E., 1981, Seasonal transformations and movements of iron in a productive English lake with deep-water anoxia, Schweiz. Z. Hydrol., 42:196.
Forbes, E.A., Posner, A.M. and Quirk, J.P., 1976, The specific adsorption of divalent Cd, Co, Cu, Pb and Zn on goethite, J. Soil Sci., 27:154.
Gadde, R.R. and Laitinen, H.A., 1974, Studies of heavy metal adsorption by hydrous iron and manganese oxides, Anal. Chem., 46:2022.
Gjessing, E.T., 1976, Physical and Chemical Characteristics of Aquatic Humus, Ann Arbor.

Gregory, J., 1978, Effect of polymers on colloid stability, in: "The Scientific Basis of Flocculation" K.J. Ives, ed., Sijthoff and Noordhoff, Alphen aan den Rijn.

Hingston, F.J., Atkinson, R.J., Posner, A.M. and Quirk, J.P., 1968, Specific adsorption of anions on goethite, Trans. 9th Int. Congr. Soil Sci. 1:669.

Hunter, K.A., 1980, Microelectrophoretic properties of natural surface-active organic matter in coastal seawater, Limnol. Oceanogr., 25:807.

Hunter, K.A. and Liss, P.S., 1979, The surface charge of suspended particles in estuarine and coastal waters, Nature, 282:823.

Jackson, T.A., 1975, Humic matter in natural waters and sediments, Soil Sci., 119:56.

Jenne, E.A., 1968, Controls on Mn, Fe, Co, Ni, Cu and Zn concentrations in soils and water : the significant role of hydrous Mn and Fe oxides. in Inorganics in Water, Adv. Chem. Ser. No. 73 pp. 337-387. Ann. Chem. Soc.

Loeb, G.I. and Neihof, R.A., 1975, Marine conditioning films, in: Applied Chemistry at Protein Interfaces, R.E. Baier ed., pp. 319-335. Am. Chem. Soc.

Lyklema, J., 1978, Surface chemistry of colloids in connection with stability, in: The Scientific Basis of Flocculation, K.J. Ives ed., pp. 3-36. Sijthoff and Noordhoff, Alphen aan den Rijn.

Mott, C.J.B., 1979, Sorption of anions by soils, in: Sorption and Transport Processes in Soils, pp. 40-53. S.C.I. Monograph No. 37.

Neihof, R.A. and Loeb, G.I., 1972, The surface charge of particulate matter in seawater, Limnol. Oceanogr. 17:7.

Neihof, R.A. and Loeb, G.I., 1974, Dissolved organic matter in seawater and the electric charge of immersed surfaces, J. Mar. Res. 32:5.

Ottewill, R.H. and Holloway, L.R., 1975, Electrokinetic properties of particles. in: "Dahlem Workshop on the Nature of Seawater", Goldberg E.D. ed. pp. 599-621.

Parfitt, R.L., Fraser, A.R. and Farmer, V.C., 1977, Adsorption on hydrous oxides. III. Fulvic and humic acid on goethite, gibbsite and imogolite, J. Soil Sci., 28:289.

Rubio, J. and Matijevic, E., 1979, Interactions of metal hydrous oxides with chelating agents 1. β-FeOOH-EDTA, J. Coll. Int. Sci., 68:408.

Ryden, J.C., McLaughlin, J.R. and Syers, J.K., 1977, Mechanisms of phosphate sorption by soils and hydrous ferric oxide gel, J. Soil Sci., 28:72.

Schnitzer, M. and Hansen, E.H., 1970, Organo-metallic interactions in soils: 8. An evaluation of methods for the determination of stability constants of metal-fulvic acid complexes, Soil Sci., 109:333.

Schnitzer, M. and Khan, S.U., 1972, "Humic Substances in the Environment", Marcel Dekker, New York.

Shaw, D.J., 1969, "Electrophoresis", Academic Press.
Tipping, E., 1981a, The adsorption of aquatic humic substances by iron oxides, Geochim. Cosmochim. Acta, 45:191.
Tipping, E., 1981b, Adsorption by goethite (α-FeOOH) of humic substances from three different lakes, Chem. Geol., 33:81.
Tipping, E. and Cooke, D., 1982, The effects of adsorbed humic substances on the surface charge of goethite (α-FeOOH) in freshwaters, Geochim. Cosmochim. Acta, 46:75.
Tipping, E., Woof, C. and Cooke, D., 1981, Iron oxide from a seasonally anoxic lake, Geochim. Cosmochim. Acta, 45:1411.
Verwey, E.J.W. and Overbeek, J.T.G., 1948, "Theory of the Stability of Lyophobic Colloids", Elsevier.

SAMPLE PRETREATMENT AND SIZE ANALYSIS OF POORLY-SORTED COHESIVE
SEDIMENTS BY SIEVE AND ELECTRONIC PARTICLE COUNTER

D.I. Little, M.F. Staggs and S.S.C. Woodman

Field Studies Council Oil Pollution Research Unit
Orielton Field Centre, Pembroke, Dyfed, UK

INTRODUCTION

Electrical sensing-zone particle counters are being used increasingly by sedimentologists for the particle size analysis of both suspended and deposited sediments, and also for the study of flocculation. Flocculation studies, however, are hampered by the effect of aggregate porosity on counter response and suitable corrections for this problem should be made (Treweek and Morgan, 1977). Several comparisons with the standard pipette method (British Standard 1377, 1975) have been made (e.g. Shideler, 1976; Behrens, 1978). Techniques using older and more recent instruments, usually based on the analysis of very small samples, have been presented and evaluated (e.g. Sheldon and Parsons, 1967; Walker et al., 1974; Dudley, 1977). Comparatively little has been written about the problems of sample pretreatment and particle size analysis of deposited cohesive sediments using the relatively low-cost Coulter Counter® Model D Industrial. This stems partly from the fact that some controversy still surrounds the use of electronic particle counters in sedimentology. This controversy relates mainly to the pretreatment of the sample and the artificial nature of the experimental system, both of which substantially alter the aggregated character of the mud. However, for the purposes of routine sediment mapping and the use of sediment textural data as a physical variable in biological monitoring, there is currently no viable substitute for the rapid analysis of the mud fraction which does not risk sampling or dispersion errors of an equal or possibly greater magnitude.

The Coulter Counter® Model D Industrial is an equally accurate but much simplified version of the more expensive particle counters in that it has a single threshold control instead of dual thresholds or a multiple channel system. It has no inbuilt data reduction facilities and a slightly narrower analytical range. These features of the instrument may represent positive advantages. Firstly, for many cohesive sediments the instrument's coarse limit of detection of approximately 75 μm necessitates the dovetailing of electronic with sieve or settling tube analyses. Provided there is no shortage of sample, the latter techniques are probably superior for the analysis of coarse material. Secondly, the manual logging of data from the digital readout permits the reduction and processing of data in any way adopted by the operator, including automatic combination with sieve analysis data. In addition, it should be remembered that whilst this paper stresses the limitations of the methods, many of the problems of this technique apply equally to alternative techniques, some of which are more time-consuming and probably, therefore, more expensive in the long term.

Operation of the Coulter Counter® Model D Industrial is based on electrical conductivity differences between sediment particles which are good insulators and a suspending saline solution which is a good conductor. A mercury manometer is unbalanced by a controlled vacuum and as it siphons back between calibrated points, draws an aliquot from the sample beaker through a small aperture across which an electrical field has been established. Unlike other Coulter Counters® this instrument does not employ a constant current device so the electrolyte (saline solution) should be of uniform temperature and ionic strength. Thus, the machine counts all the particles as they momentarily increase the electrical resistance across the electrodes, if the amplified voltage pulses so produced exceed a series of calibrated thresholds. The size of the pulse is proportional to particle volume, and the threshold calibrations are in equivalent volume spherical diameters (i.e. the diameter of a sphere of equal volume to the particle). The thresholds are adjusted for each size class required by different combinations of aperture current, amplification and threshold settings. This procedure produces cumulative oversize number frequency data which, by assuming that particles are of equal density throughout the size range, may be converted to equivalent weight percentage data for combination with the sieve results.

The development of sample pretreatment and analytical techniques using the Coulter Counter® Model D Industrial are illustrated with data from the analysis of ca. 1,000 samples. Many of these were poorly-sorted mixtures of mud, sand and gravel. Although more sophisticated particle counters are capable of sizing medium or even coarse sands (McCave and Jarvis, 1973), the analysis of muddy shell sands and muddy gravels still necessitates the dovetailing of at least two different techniques. The complete size distributions were obtained by combination of the Coulter Counter® with sieve analysis

SAMPLE PRETREATMENT AND SIZE ANALYSIS

data. The methods of combination and of data reduction are shown to influence the results.

METHODS

The phi (ϕ) scale (where $\phi = -\log_2$ particle diameter in mm) was adopted throughout the analyses. The axes on Fig. 3 show a conversion scale for ϕ. Gravel is understood to be material coarser than -1 ϕ (2.0 mm) in diameter, sand to be between -1 ϕ and 4.0 ϕ (2.0 mm and 0.063 mm), and mud to be less than 4.0 ϕ (0.063 mm). The routine methods adopted are summarised in the flowcharts (Figs. 1 and 2). Although dealt with separately below, the methods of analysis for the different size fractions cannot really be separated, because the data sets have to be related to each other. This involves determination of the sand/mud ratio from a single subsample (Fig. 1, columns 2 and 3). The mud fraction was analysed using a different subsample (Fig. 1, column 1).

Field Sampling

In intertidal areas deposited sediments were sampled by corer and a corer or grab sampler was used in subtidal areas. Corer internal diameter was ca. 6.5 cm and grab volume ca. 5 litres. Cores were sectioned and subsampled, while grab samples were subsampled to ca. 10 cm depth. The material from the latter was homogenised by thorough mixing. In all surveys, replicate samples were taken in the field to assess within-station variability. Analytical precision was assessed by laboratory subsampling and repeat analyses.

Mud Fraction

The samples were stored in a field-moist condition in plastic containers, mixed thoroughly and ca. 1 g (wet) subsampled by spatula. The removal of organic material from the sample is critical in electronic analyses which measure particle volume because organic material has substantial volume compared to its negligible weight. Removal of the organic fraction was achieved by wet-oxidation; 10-20 ml of a 6% w/v hydrogen peroxide solution was added by pipette and the sample agitated by a jet of distilled water from a wash bottle. When vigorous frothing had subsided, the samples were heated in an oven at 80°C. Further small quantities of peroxide were added and the sample bottles agitated until no further reaction was observed, usually after about seven days. The samples were then washed through a 63 µm sieve and the washings collected in a clean tray. The mud suspension was bottled, allowed to settle, the clear, supernatant peroxide solution decanted and the sample stored in glass-distilled

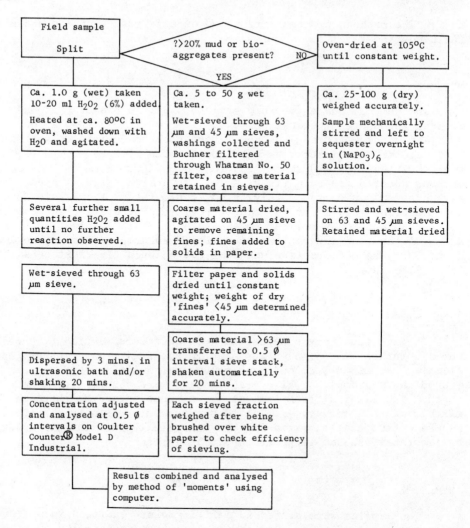

Fig. 1. Flowchart of sediment grain size analysis.

water. Samples were either analysed immediately or preserved for later analysis using a small quantity of 5% formalin to prevent micro-organism growth. The formalin was buffered to prevent loss of fine carbonate material.

Immediately prior to analysis, clear supernatant liquid in the sample bottles was decanted and the sample carefully transferred to a round-bottomed centrifuge tube to facilitate effective mixing. The sample was dispersed by three minutes' agitation in a low-powered ultrasonic bath, 55 watts power (or, in earlier samples, by

20 minutes' mechanical shaking). The required particle concentration for analysis was achieved by subsampling from the centrifuge tube using a Pasteur pipette, adding sample dropwise to a beaker containing a filtered 1% NaCl aqueous solution (Isoton® II, Coulter Electronics Ltd.). In this manner, the particle concentration was always maintained at a level well below the recommended 10% coincidence (the probability of two particles crossing the aperture simultaneously). The sample beaker was round-bottomed and fitted with a glass baffle. Each sample was made up to an approximately constant volume with clean electrolyte (Isoton® II) and constantly agitated at a high speed using the instrument stirrer mechanism. Approximately 25 ml Analar glycerol was added to the beaker to minimise the settling of large particles during analysis. These precautions help to minimise the risk of selective counting.

The 140 μm aperture tube used throughout the analyses was calibrated using 18.26 μm diameter latex spheres with instrument settings derived for 0.5 ∅ intervals. The so-called half-count technique was used (Coulter Electronics Ltd., 1976). During analysis a mean value calculated from two to eight counts was recorded for each size class from 4.0 ∅ (63 μm) down to 8.5 ∅ (2.8 μm) depending on the size and reproducibility of the counts. Background counts on clean electrolyte and glycerol were made at each size level. Previous checks on the effects of small quantities of other reagents such as formalin had been made. Checks on the stability of the dispersion were made by comparing sample concentrations before and after the analysis at the fine end of the distribution. The most significant point in the analysis of the mud fraction is that the mud was sampled for Coulter® analysis without previous drying.

Sand and Gravel Fractions

To be representative, a much larger subsample was required for the analysis of the gravel and/or sand fractions by sieving. Sediment fractions containing particles of from 2 to 20 mm diameter, if present in quantity, were sampled representatively by taking ca. 1-2 kg and analysing all the coarse material. Material coarser than 20 mm would not normally be core-sampled, although efficient sampling and dry-sieving of material of up to ca. 45 mm diameter may be attempted by analysing the entire contents of the grab. In dealing with these large samples, it is first necessary to subsample ca. 1 g for the mud analysis, and then ca. 50 g for the sand, having determined the total dry mass of the bulk sample. The method is illustrated in the flow-chart (Fig. 2) and involves dry sieving of the entire gravel fraction.

One of two procedures was then adopted for the treatment of the sand/mud fractions. Some description of the mud fraction is necessary here because the sand/mud ratio has to be determined from a single

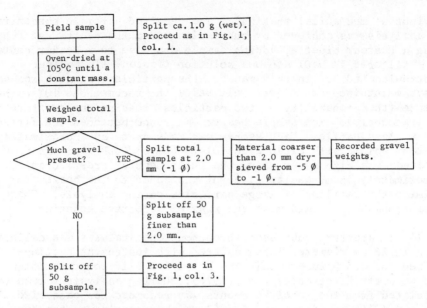

Fig. 2. Supplementary flowchart of grain size analysis for use with coarser sediments.

sample. If the samples contained no more than ca. 20% mud and if this was largely coarse silt with few bio-aggregates, then they were oven-dried to a constant mass at 105°C, weighed to 0.01 g, mechanically stirred, and left to sequester overnight in a sodium hexametaphosphate solution (3.3%). This procedure is designed to assist dispersion of individual grains. The samples were stirred again and then wet-sieved on nested 4.0 ∅ (63 µm) and 4.5 ∅ (45 µm) test sieves. Retained material was oven-dried on the respective sieve meshes. The material retained by the 4.0 ∅ sieve was then transferred to a stack of sieves ranging in 0.5 ∅ size intervals from -5.0 ∅ to 4.0 ∅. After 20 minutes of automatic shaking to fractionate the sand fraction and, if gravel was present, an additional 20 minutes for the gravel fraction, the resulting 0.5 ∅ classes were weighed. Material finer than 4.0 ∅ was added to the contents of the 4.5 ∅ sieve and this was gently brushed and agitated to remove any material finer than 4.5 ∅. The remaining material in the 4.5 ∅ class was weighed, and the total mass of material finer than 4.5 ∅ obtained by difference.

If the field samples contained a large proportion (>ca. 20%) of fine or pelloidal mud, then from 5 to 50 g (dry mass) was analysed without being initially dried and weighed. Instead, the sample was wet-sieved through 4.0 ∅ and 4.5 ∅ sieves. The washings were collected and vacuum-filtered in a Buchner funnel through a pre-weighed Whatman No. 50 filter paper. This effectively retrieved the mud from suspension. The total mass of the mud fraction could thus be determined after drying and reweighing of the filter paper and solids. Material retained on the sieves was dried, sieved and weighed as in the previous method. In both cases the mud fraction separated here is only for determination of the percentage 'fines' and is subsequently discarded.

Data Analysis

The raw cumulative particle frequency data produced by the Coulter Counter® were averaged, integrated and reduced using the technique outlined in the Instrument Manual to give equivalent weight percentages in each 0.5 ∅ class. The data reduction technique used was the full method, and included corrections for coincidence error, background counts and an extrapolation technique (Coulter Electronics Ltd., 1976). These data were further processed by being multiplied by the total mud weight (whether obtained by difference or by direct measurement after filtering) to permit their combination with the sand fraction data. Further multiplication to convert the sand and mud frequencies up to the bulk sample mass was necessary for those samples where a large quantity of gravel had been analysed and <100 g sand and mud subsampled.

The combined frequency data were analysed by the method of moments (Krumbein, 1936). The moment measures define the particle size distributions in terms of ∅ mean, ∅ standard deviation, ∅ skewness and ∅ kurtosis. The ∅ mean is the average particle diameter (and coincides quite closely with median and mode in near-normal distributions). The ∅ standard deviation is a measure of dispersion of the particle diameters about the mean size. A low value indicates that the ∅ mean is highly representative of the whole distribution and thus also indicates good sorting. ∅ skewness is a measure of symmetry about the ∅ mean. A negative sign indicates the dominance of a coarse 'tail' and a positive sign the dominance of a fine 'tail'. ∅ kurtosis measures the 'peakedness' of the size distributions. High values tend to be produced by highly peaked distributions while low values are produced by polymodal or saddle-shaped distributions.

Experimental Design

The purpose of this study is to evaluate the techniques described above by presenting and discussing the results of experiments designed to explore potential sources of error. Many of the pitfalls of particle size analysis have been fully discussed by other workers and some of their work is reviewed here. Four additional groups of problems are treated experimentally in greater detail (see Table 1).

Table 1. Summary of experimental design and statistical results.

	Objectives	Samples	Data tested	Test(s)	Results and/or decision	Significance level
①	Four separate assessments of analytical precision. H_0 = there is no significant difference between the distributions. H_a = the populations represented by the samples are different.	5 replicate field samples (pelloidal muds)	Complete % cuml. frequency ($-2\phi \to 9\phi$)	Kolmogorov-Smirnov	All 10 permutations, accept H_0	No significant differences
		3 replicate laboratory subsamples (mud fraction from muddy sand)	Mean Coulter® counts on mud fraction ($5\phi \to 8.5\phi$)	Kolmogorov-Smirnov	All 3 permutations, accept H_0	None.
		3 replicate laboratory subsamples as above (mud fraction)	Mud fraction converted to % cuml. frequency ($5\phi \to 9\phi$)	Kolmogorov-Smirnov	All 3 permutations, accept H_0	None
		3 replicate laboratory subsamples as above (sand and mud fractions)	Mud fraction converted and combined with sand data ($-1\phi \to 9\phi$)	Kolmogorov-Smirnov	All 3 permutations, accept H_0	None
②	Effects of digestion and removal of organic material from sand fraction. H_0 = organic digestion results in no significant difference between distributions. H_a = the populations represented by the samples after digestion are different.	12 pairs mixed muddy sands and gravels analysed separately, 1 with and 1 without digestion of sand fraction (organics removed from mud fractions in both). NB. The former was to B.S. 1377 and involved drying the silt/clay fraction.	Pairs of ϕ mean, ϕ standard dev., ϕ skewness, ϕ kurtosis, % fines and % gravel.	Wilcoxon Matched Pairs	ϕ mean T=30-accept H_0 ϕ std. dev. T= 0-reject H_0 ϕ skewness T=24-accept H_0 ϕ kurtosis T=10-reject H_0 % fines T=23-accept H_0 % gravel T= 8-reject H_0	None Significant at $p <0.002$ None Significant at $p <0.05$ None Significant at $p <0.02$
		11 pairs as above but ignoring gravel in all cases.	Complete % cuml. frequency.	Kolmogorov-Smirnov	All 12 pairs- accept H_0	None
			Complete % cuml. frequency $>-1\phi$.	Kolmogorov-Smirnov	All 11 pairs- accept H_0	None
③	Variations produced by combination of Coulter® and sieve data at different points. H_0 = shifting the analytical break causes no significant differences in distributions. H_a = the populations represented by the samples after shifting the analytical break are different.	20 pairs muddy sands, 1 of each with Coulter® and sieve data juxtaposed at 4ϕ, the other at 4.5ϕ.	Pairs of ϕ mean, ϕ standard deviation, ϕ skewness and ϕ kurtosis.	Wilcoxon Matched Pairs	ϕ mean T= 0-reject H_0 ϕ std. dev. T= 0-reject H_0 ϕ skewness T= 7-reject H_0 ϕ kurtosis T=87-accept H_0	Significant at $p <0.002$ Significant at $p <0.002$ Significant at $p <0.002$ None.
			% differences in $4,5\phi$ class.	Kolmogorov-Smirnov	$4,5\phi$ class T= 3-reject H_0	Significant at $p <0.002$
			Complete % cuml. frequency.	Kolmogorov-Smirnov	All 20 pairs- accept H_0	None.
		13 pairs fine muds as above.	Pairs of ϕ mean, ϕ standard deviation, ϕ skewness and ϕ kurtosis.	Wilcoxon Matched Pairs	ϕ mean T= 5-reject H_0 ϕ std. dev. T=30.5-acc. H_0 ϕ skewness T= 3-reject H_0 ϕ kurtosis T=31.5-acc. H_0	Significant at $p <0.02$ None. Significant at $p <0.002$ None.
			% differences in $4,5\phi$ class.	Wilcoxon Matched Pairs	$4,5\phi$ class T=13-reject H_0	Significant at $p >0.05$
			Complete % cuml. frequency.	Kolmogorov-Smirnov	All 13 pairs- accept H_0	None.
④	Comparison of effects of extrapolation and omission of medium and fine clay ($>8.5\phi$). H_0 = there is no significant difference between the distributions if the extrapolation procedure is excluded. H_a = the populations represented by the samples after omission of medium and fine clays are different.	11 pairs mixed muddy sands and gravels, 1 of each including and 1 excluding the extrapolation technique.	Pairs of ϕ mean, ϕ standard deviation, ϕ skewness and ϕ kurtosis.	Wilcoxon Matched Pairs	ϕ mean T= 1-reject H_0 ϕ std. dev. T= 0-reject H_0 ϕ skewness T=11-reject H_0 ϕ kurtosis T=30-accept H_0	Significant at $p <0.002$ Significant at $p <0.02$ Significant at $p <0.05$ None.
			Complete % cuml. frequency.	Kolmogorov-Smirnov	All 11 pairs- accept H_0	None.
		12 pairs very fine mud, as above.	Pairs of ϕ mean, ϕ standard deviation, ϕ skewness and ϕ kurtosis.	Wilcoxon Matched Pairs	ϕ mean T= 0-reject H_0 ϕ std. dev. T= 0-reject H_0 ϕ skewness T= 9-reject H_0 ϕ kurtosis T= 4-reject H_0	Significant at $p <0.002$ Significant at $p <0.02$ Significant at $p <0.02$ Significant at $p <0.02$
			Complete % cuml. frequency.	Kolmogorov-Smirnov	4 out of 12 prs.-reject H_0	Significant at $p <0.10$ (3 prs.) and $p <0.05$ (1 pr.)

These are:

1. The assessment of analytical precision, including field replication, laboratory replication, and the effects of conversion to percentage frequency by mass on analytical repeatability. (Results are presented in section 3.1 and discussed in 4.1.1 and 4.1.2.)

2. The effects of the removal of organic material from the sand fraction on the size distribution of the mud fraction and on the overall distribution. (Results are presented in section 3.2 and discussed in 4.1.4.)

3. The effects of combining sieve and particle counter data at different points. (Results are presented in section 3.3 and discussed in 4.3.1.)

4. The comparison of grain size distributions with and without an extrapolation technique included in the data reduction routines. (Results are presented in section 3.4 and discussed in 4.3.2.)

These experiments were carried out on a range of survey data collected from estuarine areas where both mixed and better-sorted cohesive sediments were present. Because the aims were to compare alternative experimental methods, it was anticipated that analysis of variance techniques would be used. There were, however, several reservations about the applicability of these powerful parametric tests to the data. These included serious departures of the experimental data from the normal distribution and, in one case, doubts that the observations were completely independent. In spite of the fact that the non-normal data may be transformed and that the sample variances were generally homogeneous, it was considered prudent to employ non-parametric tests throughout.

With each group of experiments (Table 1, 1 to 4) the null hypothesis was that there is no significant difference between distributions. Two approaches to testing these hypotheses were used: the pairs of entire cumulative percentage frequency distributions were compared for differences using the Kolmogorov-Smirnov Test (Till, 1974). Differences in the moment parameters, themselves sensitive to the entire distribution, were tested using the Wilcoxon Matched Pairs Test (Meddis, 1975). The latter test was also used to focus attention on some size classes where experimental alterations in methodology were expected to produce differences. All tests were two-tailed because it could not be correctly predicted which direction of change would result from an experimental manipulation of the data.

RESULTS

Some 120 statistical tests were performed on the results of the four groups of experiments. The results are summarised in Table 1 together with their significance levels and decisions regarding the null hypotheses. It should be noted that, with the Kolmogorov-Smirnov Test, to suspect with 95% confidence (p <0.05) that a pair of samples were from different populations, the largest absolute difference in any one size class between the two cumulative percentage frequency distributions would have to be at least 19.23%. To be 90% confident (p <0.1) the difference would only have to be 17.25%.

Analytical Precision

The variability of field samples in very pebbly cohesive sediments can be problematical because of the difficulties of obtaining a large enough sample. Tables 1 and 2(a) and Fig. 3, however, show that highly reproducible results can be obtained in slightly better-sorted deposits. The largest difference in percentage frequency for any class pair on the cumulative curve was 6.96% which, using the Kolmogorov-Smirnov Test, was concluded to be insignificant. Laboratory precision was equally good, even for replicate subsamples where there was very little fine material (<5%). Table 2(b) shows that mean particle counts on the mud fraction were also very reproducible. The highest class difference in percentage frequency here was only 0.36% (in the 7.5 ∅ class). The Kolmogorov-Smirnov Tests were repeated to check reproducibility after conversion of the raw mud number frequencies to a mass frequency distribution, by multiplying them by the total mass of silt and clay. This time the largest differences were in the 5.0 ∅ and 5.5 ∅ classes, but they were still not significant (3.08 to 6.22%). All these tests indicated that analytical precision of the mud data was good. Nor were any differences significant when the converted mud fraction data were combined with the sand fraction data. It was interesting to note that the largest differences in cumulative curves then occurred in the 3.0 ∅ class (7.12%).

The Effects of Removal of Organics from the Sand Fraction

The somewhat conflicting results of the second group of experiments are shown in Table 1 and a representative pair of grain size histograms is illustrated in Fig. 4. Kolmogorov-Smirnov Tests of the overall distribution showed that removal of the organic material does not significantly alter the cumulative curve (largest class differences ranged from 4.17 to 15.35%). The Wilcoxon Matched Pairs Test, however, indicated that the digestion process tended to produce

Table 2(a). Summary of grain size parameters from the analysis of five replicate grab samples from one station (estimate of field variability).

	$\emptyset$ mean	$\emptyset$ standard deviation	$\emptyset$ skewness	$\emptyset$ kurtosis	% <63 µm
Sample 1	2.17	2.50	0.96	3.36	20.40
2	2.31	2.61	0.86	3.04	22.94
3	2.58	2.62	0.92	3.02	22.05
4	2.37	2.35	0.91	3.38	23.26
5	2.42	2.51	0.91	3.10	23.39

Table 2(b). Mean counts (background count deducted) from the analysis of three replicate subsamples from a single field sample (estimate of laboratory precision).

µm diameter	$\emptyset$ diameter	Subsample (a)	Subsample (b)	Subsample (c)
31.3	5.0	3.0	3.25	1.5
22.0	5.5	11.6	18.6	11.9
15.6	6.0	38.8	42.6	40.7
11.0	6.5	110.5	111.5	124.0
7.8	7.0	384.0	409.0	393.0
5.5	7.5	1392.0	1476.0	1459.0
3.9	8.0	4712.0	4767.0	4884.0
2.8	8.5	16353.0	16667.0	16744.0

significant differences in $\emptyset$ standard deviation, $\emptyset$ kurtosis and percentage gravel. The trends to finer $\emptyset$ mean, more negative $\emptyset$ skewness and a lower percentage fines were not significant, although visible in the majority of pairs. The significant change in percentage gravel ($p < 0.02$) was unexpected because the same gravel raw data had been used. To check whether the large quantity of gravel in some samples had obscured the result from the Kolmogorov-Smirnov Tests, they were then repeated ignoring the gravel fraction. The results still indicated no significant differences although many approached the 90% confidence limit, the largest difference being 15.48%. The differences show, however, that the pretreatment of material outside the analytical range of the Coulter Counter® can cause appreciable alterations to the combined data.

Variations Produced by Combination of Coulter® and Sieve Data at Different Points

The objective of the third group of experiments was to determine the optimal point at which to combine sieve and particle counter

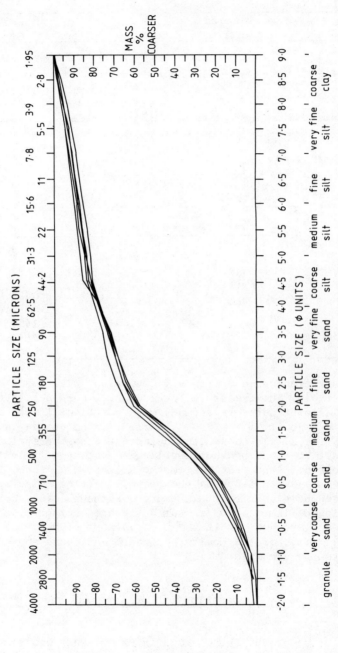

Fig. 3. Cumulative percentage mass curves for five sediment subsamples taken from replicate grabs. See Table 2(a) for summary of statistical parameters.

analyses with a minimal analytical break. The results are shown in
Table 1, and two pairs of grain size histograms produced by the two
experimental treatments are illustrated in Fig. 6. The most dramatic
differences occurred, as might be expected, in the 4.5 ∅ class.
Although ranging from negligible to over 8% of the total sample mass,
the main trend discernible from the Wilcoxon Matched Pairs Test was
a significant reduction in the 4.5 ∅ class when the Coulter Counter®
was used to measure this size level. The differences between the
electronic technique and the sieve technique were most significant
($p < 0.002$) when the area of overlap was near the sample mode, as
shown in the lower histograms in Fig. 6.

The extent to which a deficiency in one size class affected the
overall frequency distribution was assessed by a Wilcoxon Matched
Pairs Test on the moments from two sets of grain size distributions.
The results indicated that use of the Coulter Counter® data for
the 4.5 ∅ class caused a significant trend to finer ∅ mean in the
case of both the muddy sands ($p < 0.002$) and the fine muds ($p < 0.02$).
∅ standard deviations were only significantly different in the case
of muddy sands ($p < 0.002$), and ∅ skewness differences were highly
significant in both series of sediment samples ($p < 0.002$). The
∅ kurtosis changes were insignificant. As with the digestion experiments, the Kolmogorov-Smirnov Tests seemed less sensitive to changes
than the data moments tested by the Wilcoxon Matched Pairs Test. The
largest difference in the fine mud series was 13.49% and in the muddy
sand series, only 4.97%. Thus, each of the pairs of cumulative percentage distributions was drawn from the same statistical population.

Comparisons of Extrapolation and Omission of Medium to Fine Clays

The results of the fourth group of experiments are shown in
Table 1 and a pair of grain size histograms is illustrated in Fig. 7.
The results indicated that for the very fine muds, very significant
changes in the moment parameters were produced by omission of the
extrapolation routine during data reduction. The ∅ mean and ∅
standard deviation were the most affected ($p < 0.002$) with ∅ skewness
and ∅ kurtosis less so ($p < 0.02$). The Kolmogorov-Smirnov Tests
endorsed this result in three out of 12 pairs of cumulative percentage
mass distributions ($p < 0.1$) with differences of up to 20.05% ($p < 0.05$)
in one case. The errors introduced by failing to extrapolate were
less dramatic in a series of coarser, more mixed sediment samples.
Although the Kolmogorov-Smirnov Tests detected no significant difference (largest difference was 5.16%), the Wilcoxon Matched Pairs
Test showed ∅ mean ($p < 0.002$), ∅ standard deviation ($p < 0.002$) and ∅
skewness ($p < 0.05$) as being significantly different.

DISCUSSION AND EVALUATION OF TECHNIQUES

Sample Pretreatment

Field Subsampling. Avery (1974) gives a useful guide to the minimal sample sizes for estimating stone content, a factor which normally determines the degree of field subsampling required for the finer fractions. Pebbly cohesive sediments present very serious sampling problems, and many workers consequently appear to ignore the extreme coarse 'tail' (for example Kelley and McManus, 1970). Isolated pebbles, cobbles and boulders are sometimes an important microhabitat for both epilithic flora and fauna on cohesive sediments and are therefore of relevance in ecological studies. In practice, however, some size cutoff is needed, and on the basis of a five-litre grab volume, the upper limit is approximately $-5.5\ \emptyset$ (45 mm) effective particle diameter. To raise this limit to $-6.0\ \emptyset$ (64 mm) would involve the sampling and analysis of a further four grabs, which would probably be an excessive sampling effort at one station in a wide-ranging survey.

Laboratory Subsampling. Fine sediment samples may be dried or analysed wet, a choice which largely determines the sort of sample splitting which is needed. The pretreatment techniques outlined in B.S. 1377 (1975) involve air-drying of the sample, wet-oxidation in H_2O_2 (hydrogen peroxide) and subsequent oven-drying. There is serious doubt as to whether clay particles can be resuspended after drying (Sheldon and Parsons, 1967). In addition, water may be driven from the clay lattice, thus altering the particle volume. For these reasons the standard method was not followed.

An advantage of the use of the relatively small work samples required for Coulter Counter® analysis is that much of the bulk sample remains for a parallel sieve analysis. Buller and McManus (1979) considered the splitting of the sample into two equal subsamples, perhaps the best alternative to drying. In their method, "the first portion is wet-sieved, and the sediment passing the 63 μm sieve retained. Both the sand and mud fractions are dried and weighed in order that the relative proportions of the two parts may be established." The second portion is also wet-sieved and the sand retained, dried and weighed, while the washings containing the mud fraction are collected and analysed. The weight of the mud is thus determined indirectly. This is essentially similar to the procedure advocated herein for samples which should not be dried at the outset. Buller and McManus (1979) imply, however, that a larger sample of fine mud is needed, perhaps for pipette or hydrometer analysis.

In the present method, the small size of the subsample needed for Coulter Counter® analysis necessitates very thorough mixing of the bulk sample in its container. This homogenisation can be very effective with liquid mud and muddy sands, and highly reproducible results have been obtained (see Fig. 3, Tables 1, 2(a) and (b)). Another advantage of subsampling from the bulk sample and not after wet-sieving is that the digestion process can be commenced several days before the analysis (see Fig. 1). This not only allows time for complete oxidation of the organic material but also permits the mud sample preparation to be separated from any dry-sieving procedures, an advantage in terms of operator convenience and the avoidance of dust contamination.

It must be emphasised that the riffle-box sample splitters recommended in B.S. 1377 and many other commercially-available splitters only work efficiently on dry sediment and even then, particles appear to be lost to the air and to the metal surfaces of the riffle box. The cone and quartering method of subsampling is practical only for larger, more equal-sized subsamples, and is again most appropriate for dry powders. The importance of obtaining a representative subsample by a standardised technique is emphasised by Shideler (1976). His method involves agitation of the concentrated sample in a beaker equippped with a baffle to disrupt vortex motion and prevent hydraulic fractionation. The subsample is then obtained by automatic pipette from top, middle and bottom of the beaker. These precautions would have to be borne in mind by workers adopting any subsampling technique after the wet-sieving stage which, because of the relatively large quantity of water involved, necessitates subsampling from a diluted suspension. The point is to avoid the settling of large particles.

The representativeness of sample splitting can largely determine the reproducibility of the technique. From the results of the first group of experiments, however, it was also evident that the reproducibility of the technique was worse when the Coulter Counter® data from the mud fraction were converted to percentage mass, and then multiplied by the percentage mud for combination with sieve data. Data reduction of this kind is necessary with most poorly-sorted cohesive sediment samples. Caution must, therefore, be exercised when converting to percentage mass, especially when particles of widely different specific gravities are analysed at intervals of 0.5 ∅ or greater. However, some of the error probably originated not with the Coulter Counter® analysis but with wet-sieving at 4.0 ∅ and the subsequent dry-sieving. Although these grain sizes are beyond the detection limits of the Coulter Counter® Model D Industrial, their imprecision might reduce overall confidence in the data. In summary, both splitting and data reduction are considered as important variables in improving analytical precision.

Treatment of Organic Material. Another problem of sample preparation for combined sieve and electronic analysis is whether or not to remove all organic material (as recommended in B.S. 1377) or to remove it from just the mud fraction. Experiments designed to elucidate this problem (group 2) showed that the only significant differences occurred in ∅ standard deviation, ∅ kurtosis and percentage gravel after oxidation in H_2O_2. Most of the changes can be explained by the removal of organic material from the very fine sand grade (3.0 - 4.0 ∅) accompanied by the detection of larger quantities of very fine silt and coarse clay (>7.0 ∅) (see Fig. 4). It is possible that organic material was present in sand grade bio-aggregates and that associated fine mineral particles were released from these aggregates during digestion. They would then be free to pass the 4.0 ∅ sieve and thus be included in the Coulter® analysis. It should be noted that this experiment effectively compared the adopted method with that of B.S. 1377 in terms not only of oxidation of organic matter but also of subsequent drying and weighing of the silt/clay fraction. As a result, comparison of the errors involved included errors caused by changes in the silt and clay due to oven-drying.

It is, perhaps, surprising that the differences were not much more significant. From loss of weight on ignition analyses, however, the mass of organic matter rarely exceeds 10% of the sample, and this figure usually correlates closely with the percentage mud. Since the organic material is routinely removed from the mud subsample, the potential error is thus not as great as might be expected. Although it is an undesirable element in the pretreatment procedure, the reason for removing organic material from the mud fraction is quite plainly that Coulter Counter® or, indeed, pipette measurement of particle size would be seriously affected. No such justification can be made for digestion in the sand size range and since it is extremely time-consuming and does not, in most cases, dramatically alter the result, it is considered unnecessary. In cases where either the gravel content or the organic content is elevated, however, it may be considered necessary to remove all the organics to standardise treatment of both sand and mud fractions. This is because in pebbly sediments which have been subsampled, the sand/mud frequency data are commonly multiplied up to a bulk sample mass, and if the organics are not removed the percentage frequency of the gravel is significantly diminished ($p < 0.02$). This occurred in spite of the fact that identical gravel data were used in the experiment (see Fig. 4).

Screening. The removal of organic material from the mud subsample is normally followed by screening of the sample, conventionally at 63 μm (4.0 ∅) to separate sand from mud and reduce the risk of aperture blockage. If a larger sand and gravel sample is being analysed in parallel, then the material coarser than 63 μm in the subsample used for mud analysis may be discarded. If not, then

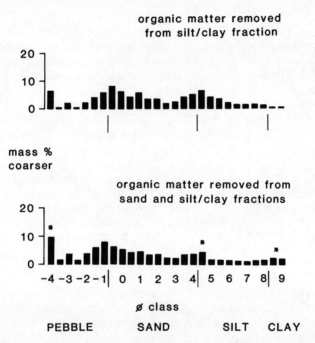

Fig. 4. Grain size histograms showing the effects of removing organic matter from the sand fraction in addition to the mud fraction. (* indicates main differences.)

there is a coarse tail which remains unanalysed unless a two-tube technique is adopted (see section 4.2.1). It is possible to wash the sieve mesh with a jet of distilled water and still minimise the quantity of the washings, which are collected and subsequently analysed. In spite of the necessity of this screening procedure to prevent aperture blockage, it has been observed that wet sieving usually leaves some mesh-sized particles trapped in the film of water on the sieve. It is possible that this is a contributing factor to the apparent deficiency of particles in the size class immediately below 63 µm, a reported feature of many analyses involving two techniques (Buller and McManus, 1979). This point is taken up more fully in section 4.3.1.

Dispersion. One of the most important aspects of pretreatment is the dispersion of the stock sample. In this connection, it is a positive advantage of the electrical sensing zone method that periodic checks for flocculation and deflocculation under certain circumstances can be made during an actual analysis without taking a subsample for microscopic examination (Table 13, Instrument Manual, Coulter Electronics Ltd., 1976). Stirring techniques, changes of electrolyte and ultrasonics were the three approaches to dispersion evaluated.

The majority of samples can be dealt with by stirring and chemical dispersion procedures. The use of sodium hexametaphosphate solutions (B.S. 1377, 1975) and of Nonidet P42 (Dudley, 1977) have been recommended in conjunction with overnight shaking. However, for some trial samples where mechanical stirrers and shakers were used it was found that a procedure which improved dispersion stability in one sample discouraged it in another. It is probable that the greater turbulence involved in such methods promotes inter-particle collisions, although this orthokinetic flocculation does not necessarily alter the stability of the system. The equilibrium between floc growth and fluid shear has been described by Zeichner and Schowalter (1977), and is related to the input of mechanical energy, a parameter which is difficult to quantify.

Alternative approaches to sample dispersion also include a change of electrolyte, for example from a sodium chloride to a tri-sodium orthophosphate solution at a 4% ionic strength. This has proved successful in stabilising micritised carbonate samples. Changing electrolytes is a time-consuming chore on a sample-to-sample basis, however, because the whole glassware assembly has to be drained each time. Because of the high stirrer speeds used during Coulter® analysis, no attempt has been made to preserve the aggregated character of the mud fraction. Until alternative methods are evaluated, dispersion by ultrasonics is thought to be a repeatable compromise, although it should be noted that ultrasonic probes on their lowest power settings have been observed to possibly cause

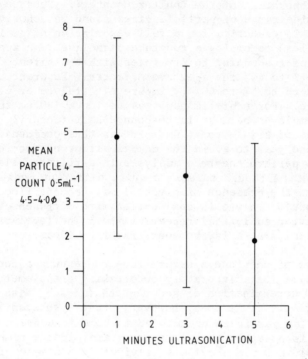

Fig. 5. The effects of the use of an ultrasonic probe (65 watts) for sample dispersion. Mean counts in the 4.5 - 4.0 ∅ class for three periods of ultrasonication are shown. Counts were stable at the fine end, so fall in count was possibly due to fracture of particles.

fracture of the particles after only five-minute periods of dispersion (see Fig. 5). Dispersion in a low-powered ultrasonic bath thus appears to be the most satisfactory method, permitting a more standardised approach (Sheldon and Parsons, 1967; Coulter Electronics Ltd., 1976).

Coulter Counter® Analysis

Choice of Aperture Tubes and Electrolyte. The selection of appropriate aperture tubes and electrolytes is fundamental to successful operation of the Coulter Counter®. Because of the extremely wide range of particle sizes found in cohesive sediment systems, it is essential for a full description of grain sizes down to 1 μm or less to employ a two-tube technique. An aperture tube is capable of responding to particles with equivalent diameters over a range of 2% to 40% the aperture diameter. In practice, this range may be reduced to 5% to 40% by impure electrolyte or electrical interference. For a preliminary evaluation a 140 μm tube was routinely employed because it responded satisfactorily to particles in the range of 2.8 μm to 56 μm. It was thus decided for a trial period of one year to avoid the complications of two-tube techniques, and instead perform one-tube analyses in a filtered electrolyte (Isoton® II). Although this is a quite different medium from seawater and the presence of glycerol may promote flocculation, the use of this electrolyte in conjunction with ultrasonic dispersion has facilitated analytical repeatability. The level of realism of the analytical system is, however, far from ideal.

The use of only one aperture tube represents a considerable saving of time for a laboratory undertaking granulometric studies where rapid determination of sediment textures by mass is the primary objective. It does, however, leave a substantial fine 'tail' of clay particles unanalysed in some fine grain size distributions. It is possible that an appropriate pipette sample from a settling column would provide a gravimetric estimate of this tail. Shideler (1976) has outlined a satisfactory method for conducting two-tube analyses with the more sophisticated Model TA Coulter Counter®, whereby runs of eight samples would be analysed by 200 μm tube, the samples screened at 20 μm, and then analysed again using a 30 μm tube. This procedure is effectively a compromise between the chore of constant tube-changing and the analytical imprecision which would result from instability of the dispersion if the samples were stored for longer than four hours (Shideler, 1976).

Installation and Interference. The second consideration of importance particularly to the fine end of the size distribution is the minimisation of background electrical noise, which may be caused by mains interference, impure electrolyte or radiated noise from other electrical equipment, notably gas chromatographs, oven thermostats or fluorescent lighting. A range of countermeasures is available to combat these sources of error, including mains filters, micropore electrolyte filter systems and earthed shields, but it should be noted that these can be quite expensive accessories to the basic instrument supplied by the manufacturers. Their successful implementation, however, is an insurance against the loss of laboratory time and against the frustration of the operator. The instrument should be protected from vibration or dust. Some of these points have already been made by Shideler (1976) and in the Instrument Manual (Coulter Electronics Ltd., 1976).

Some Causes of Selective Counting. Further emphasis is needed on the problem of subsampling at the analysis stage. The method evolved is believed to be a satisfactory way of ensuring repeatable subsampling from the digested and dispersed stock sample. The repeated pumping action made possible by the bulb filler mechanism randomises the subsampling procedure when used in conjunction with round-bottomed centrifuge tubes.

Shideler (1976) has addressed another problem which usually arises during the analysis, that of sample dispersion and settling in the sample beaker. He showed that different degrees of agitation produced dramatically different histograms of grain size distributions. The use of standard volumes of electrolyte and stirrer settings was recommended to minimise these errors (Shideler, 1976). In this context, it has also been found useful to use round-bottomed beakers equipped with baffles to reduce vortex motion and inhibit the introduction of air bubbles at the highest stirrer speeds. These measures, in combination with the addition of glycerol to ca. 10% v/v to increase the viscosity of the electrolyte, successfully prevented settling of mud particles up to 45 μm in diameter. Larger quantities of glycerol (up to 30% v/v) helped to maintain slightly coarser material than this in uniform suspension. However, in one series of samples with very distinctive disc-shaped particles there was an apparent tendency for them to be fractionated on the basis of their shape. When using glycerol, to avoid interference the electrolyte must contain an equal proportion of glycerol throughout the beaker, aperture tube and control piece system. Adjusting this proportion is laborious on a sample-to-sample basis because the instrument may need draining of electrolyte. A slight increase in the duration of counting must also be expected.

Data Reduction and Analysis

Combination with Sieve Data. Previous evaluation of Coulter Counter® techniques has not emphasised the problems of combining data from sieve or settling tube analyses with those from electronic analyses. In terms of repeatability and time-saving during the data reduction stage of analysis, it is considered important to select as objectively as possible the point at which to begin the electronic analysis. The group of experiments (Table 1, group 3) designed to explore the possibilities of overlapping sieve and Coulter Counter® data showed that significant differences occurred in the 4.5 ∅ class when it was analysed by the two alternative techniques. These differences affected ∅ mean, ∅ standard deviation and, in one set, ∅ skewness parameters, although changes in the overall frequency distribution were not significant as detected by the Kolmogorov-Smirnov Test. The inadequacy of the electronic technique at its coarse limit of detection means, however, that there is a real danger that bimodal size distributions will be described in error when the overlap is near the sample mode (Fig. 6). The choice of particle diameter at which to juxtapose the two techniques clearly influences the impression of the grain size distributions created by the analysis.

A range of contributing factors can be invoked to explain these results. Some have already been suggested, such as the possible loss of material in the mesh during the screening of the subsamples used for mud analysis. This would not have affected the larger subsamples used for sand analysis and determination of total mud content because the sieves were dried after wet-sieving and thus sieve aperture-sized material was dislodged into the next class. Additionally, the serious problems of maintaining a uniform suspension in the Coulter® sample beaker due to settling might have caused selective counting of material finer than 4.5 ∅. A significant proportion of the apparent deficiency in the 4.0 - 4.5 ∅ class must, however, be attributable to the 'loss' of particles at the extreme coarse limit of detection of the aperture tube used. It is also possible that the discrepancies were caused by a slight difference in the parameters measured by the two techniques: 63 µm determined by a sieve is likely to represent the particle's intermediate axis, which may be significantly smaller than the equivalent volume spherical diameter, especially in elongate particles.

Because of the present inability to differentiate between these factors, it is not yet possible to improve on the assertion that the discontinuity occurring at approximately 45 µm is an 'analytical break'. This is at variance with the findings of Belderson (1964) reviewed in Swift et al. (1972), who concluded that the break was a real feature of the frequency distributions they examined and not an analytical artifact. Buller and McManus (1979) emphasise that the analytical break occurs with other techniques. In a pipette analysis, for example, it is often difficult to take the first sample at the precise interval after shaking the suspension, and it is this sample which determines the total mass of the silt/clay fraction.

It is important, therefore, in selecting objectively the point of combination of Coulter Counter® with sieve data, that overlapping analyses be performed so that the effects of methodological errors can be properly assessed. Combination at 45 µm is one way of satisfactorily minimising the analytical break.

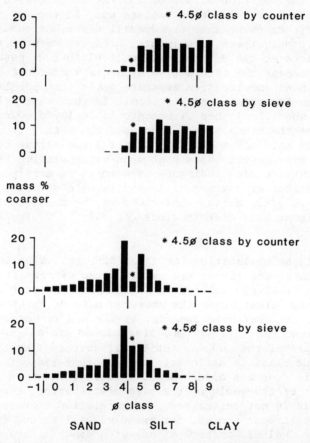

Fig. 6. Grain size histograms showing the effects of combining sieve and particle counter analyses at different points.

Extrapolation and Computerised Data Handling. Coulter Counter® data from cohesive sediments seldom close the distributions at the fine 'tail'. On account of the dangers of extrapolation when the method of moments is employed in the reduction of grain size data (Buller and McManus, 1979), it is important to consider this limitation.

The results of the extrapolation experiments (Table 1, group 4) showed that the omission of medium to fine clays caused significant changes in the ∅ grain size parameters and, in some of the very fine muds, significant changes in the overall distributions. Of all the tests carried out, these experimental differences were the most dramatic, since at the 90% confidence level in four cases it was possible to accept the alternative hypothesis that the entire distributions had been sampled from separate grain size populations (see Fig. 7). The importance of this result is that extrapolation reduces such errors when the method of moments is employed. On the other hand, because the moments technique uses the entire frequency distribution and not just a few points on the cumulative curve, confidence in the data cannot be as high when extrapolation is used for very fine silt and clay sediments as when it is merely used to close the distributions of coarser silts and sandy sediments. In this context, it is worth noting that there may be some doubt as to whether sediment size distributions can ever be truly closed at the fine end.

Part of the explanation for these differences is that when both particle counter and sieves are used, if the extreme fine end of the distribution is simply omitted by being below the particle counter's lower detection limit, then the measured mass of omitted material is crudely redistributed over the rest of the mud frequency distribution. This is because the Coulter® mud frequencies are all multiplied by the total mass of the mud, an additional source of error which also occurs if one class is deficient for any other reason, as shown in Fig. 6. This does not occur when the Coulter Counter® is used for the analysis of the whole of the size distribution, because the mass of the sample is not determined. Extrapolation therefore permits the more rigorous comparison of, for example, well-sorted fine muds (analysed by Coulter Counter® alone) with sandy or gravelly muds (analysed by both techniques). A further point concerning extrapolation is that it is highly probable that the above remarks would still apply if the more laborious two-tube technique were to be adopted routinely. The percentage error in this case, however, would be reduced because smaller aperture tubes are capable of detecting ca. 1 μm particles.

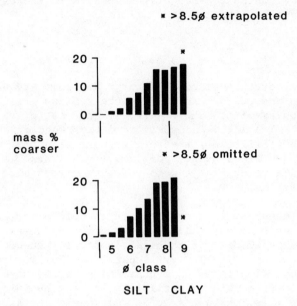

Fig. 7. Grain size histograms comparing the extrapolation and omission of the >8.5 ∅ fractions. Note: the extrapolated clays are lumped in the 9.0 ∅ class.

The reduction of Coulter Counter® data is manually laborious. The data handling and storage facilities provided by a microcomputer system are, therefore, to be strongly recommended for all aspects of establishing electronic particle counting in a sedimentology laboratory. In the case of the Coulter Counter® Model D Industrial, there is no commercially available data reduction equipment, and even where more sophisticated models are concerned, the software appears to be limited in terms of plotting composite results from two-tube analyses (Shideler, 1976). When combining sieve and Coulter Counter® data routinely, the use of a computer is highly advantageous to maintaining speed of sample analysis. This stems from the fact that very few deposited cohesive sediment systems can be measured entirely by this instrument, with or without the two-tube technique, which consequently necessitates the use of sieves for the analysis of the coarse 'tail'.

CONCLUSION

1. The electronic particle counter has some advantages of speed and reproducibility in the routine grain size analysis of mud samples. The relatively low cost of the Coulter Counter® Model D Industrial and the range of accessories which facilitate its effectiveness, mean that the technique deserves wider usage, particularly by those laboratories which already have a computer system to assist with data processing.

2. The technique is adequately reproducible for purposes of routine sediment characterisation, mapping and for the use of textural data as physical 'controls' in ecological monitoring. The latter is particularly important when the biological effects of industrial activity have to be separated from natural variability of the benthic community. To do this, there has to be reasonable certainty that the sediment type sampled on each survey is strictly comparable to that sampled on previous surveys. In some instances, ecological damage due to, for example, the oil industry might include actual changes in sediment texture associated with the incorporation of drilling muds and cuttings. These are detectable by the present methods, and would be a useful supplement to geochemical data.

3. A method of analysis which permits the combination of sieve and Coulter® data is necessary for the majority of cohesive sediments. Cohesive sediments can be representatively subsampled unless very pebbly, provided simple precautions are carefully observed. Organic material in the mud fraction has to be removed and the sample for Coulter Counter® analysis thoroughly dispersed before full repeatability can be achieved. Although this pretreatment prevents analysis of the sediment in an undisturbed state, it is a problem which faces many other particle-sizing techniques. The method presented is a satisfactory compromise. Removal of the organic matter from the sand fraction is not considered necessary for most samples.

4. Data processing is shown to be a major source of variability in some experimental data. In particular, the juxtaposition of sieve and Coulter Counter® data must include an overlap,

until the optimal analytical break point for the laboratory analysis is known. Both standard sieve and Coulter Counter® Model D Industrial methods are near their limits of detection and their errors can be misleading, especially when the area of overlap coincides with a sample mode. When subsampling from gravelly or sandy muds, errors occurring outside the limit of detection of the Coulter Counter® are shown to affect the data because of the multiplication factors necessary for data combination.

5. The extrapolation technique is recommended to close fine tails in grain size distributions, and this involves less error than omission even when the method of moments is employed. Caution in interpretation must be exercised, however, when the sediment is extremely fine. It is possible that a pipette sample to determine the silt/clay ratio would act as an independent check on the present method in such cases.

6. Other areas of this work requiring further attention include two-tube techniques, the possible minimisation of pretreatment and the analysis of the samples at 0.25 $\emptyset$ intervals. Respectively, these would facilitate an extension of the detection limit towards a modal area for number frequencies, a possibly less artificial analytical system, and more detailed grain size descriptions.

ACKNOWLEDGEMENTS

We are grateful to the staff of Coulter Electronics Ltd. for their advice and assistance at all times. The Department of Chemical Engineering, University College, Swansea, have provided useful pointers, whilst Mr. J. A. J. Mullett and Dr. J. D. D. Bishop have offered the benefits of their experience in particle counting and statistics respectively. Mr. A. Thompson and Dr. J. M. Addy assisted with the design of Basic programs and the development of computing facilities. We also gratefully acknowledge the work of Mrs. L. Evans in preparing the manuscript, and the assistance of our colleagues at the Oil Pollution Research Unit both in the field and in commenting on the manuscript.

REFERENCES

Avery, B. W., 1974, Introduction, in: "Soil Survey Laboratory Methods," Soil Survey Technical Monograph No. 6, B. W. Avery and C. L. Bascomb, eds., Rothamsted Experimental Station, Harpenden.

Behrens, E. W., 1978, Further comparisons of grain size distributions determined by electronic particle counting and pipette techniques, J. Sed. Pet., 48: 1213.

Belderson, R. H., 1964, Holocene sedimentation in the western half of the Irish Sea, Mar. Geol., 2: 147.

British Standard 1377, 1975, "Methods of Test for Soils for Civil Engineering Purposes," British Standards Institution, London.

Buller, A. T. and McManus, J., 1979, Sediment sampling and analysis, in: "Estuarine Hydrography and Sedimentation," K. R. Dyer, ed., Estuarine and Brackish Water Sciences Association, Cambridge University Press.

Coulter Electronics Ltd., 1976, "Instruction Manual for Coulter Counter® Model D (Industrial)," 4th edition, Harpenden.

Dudley, R. J., 1977, The particle size analysis of soils and its use in forensic science: the determination of particle size distributions within the silt and sand fractions, J. Forens. Sci. Soc., 16: 219.

Kelley, J. C. and McManus, D. A., 1970, Hierarchical analysis of variance of shelf sediment texture, J. Sed. Pet., 40: 1335.

Krumbein, W. C., 1936, Application of logarithmic moments to size frequency distributions of sediments. J. Sed. Pet., 6: 35.

McCave, J. M. and Jarvis, J., 1973, Use of the Model T Coulter Counter® in size analysis of fine to coarse sand, Sedimentology, 20: 305.

Meddis, R., 1975, "A Statistical Handbook for Non-Statisticians," McGraw-Hill, Maidenhead.

Sheldon, R. W. and Parsons, T. R., 1967, "A Practical Manual on the Use of the Coulter Counter® in Marine Science," Coulter Electronics Sales Co., Canada.

Shideler, G. L., 1976, A comparison of electronic particle counting and pipette techniques in routine mud analysis, J. Sed. Pet., 46: 1017.

Swift, D. J. P., Schubel, J. R. and Sheldon, R. W., 1972, Size analysis of fine-grained suspended sediments: a review. J. Sed. Pet., 42: 122.

Till, R., 1974, "Statistical Methods for the Earth Scientist," MacMillan, London.

Treweek, G. P. and Morgan, J. J., 1977, Size distributions of flocculated particles: application of electronic particle counters, Environ. Sci. and Tech., 11: 707.

Walker, P. H., Woodyer, K. D. and Hutka, J., 1974, Particle size measurements by Coulter Counter® of very small deposits and low suspended sediment concentrations in streams, J. Sed. Pet., 44: 673.

Zeichner, G. R. and Schowalter, W. R., 1977, Use of trajectory analysis to study stability of colloidal suspensions in flow fields, A.I.Ch.E. Journal, 23: 243.

SIZE DISTRIBUTIONS OF SUSPENDED MATERIAL IN THE SURFACE WATERS

OF AN ESTUARY AS MEASURED BY LASER FRAUNHOFER DIFFRACTION

A.J. Bale, A.W. Morris, and R.J.M. Howland

Institute for Marine Environmental Research
Prospect Place, The Hoe, Plymouth, Devon, UK

INTRODUCTION

Heterogeneous processes, responsive to both the abundance and quality of suspended particles, are of major importance in the chemistry of estuarine waters (Burton and Liss, 1976). The low salinity region of an estuary is often an area of particularly high chemical and biological activity (Morris et al., 1978, 1982a) and high turbidity can be a major contributory factor (Morris et al., 1982b). Studies at IMER of the physical and chemical characteristics of estuarine suspended particulates are aimed at resolving the complex heterogenous chemical and biological interactions in estuaries. These studies are concentrated on the Tamar Estuary, southwest England, for which considerable information on the chemistry is available (Butler and Tibbits, 1972; Morris, 1978; Morris et al., 1978; Loring et al., 1981; Morris et al., 1982a,b,c; Readman et al., 1982).

Recent studies of the abundance and chemical composition of suspended particles in the Tamar Estuary (Loring et al., 1981; Morris et al., 1982c) have demonstrated marked temporal and spatial variabilities induced by tidally controlled, cyclic resuspension and deposition of sediment and by selective retention of particles in suspension within the turbidity maximum zone.

Here we discuss the application of laser Fraunhofer diffraction to studies of the size characteristics of estuarine suspended particles and present some preliminary results. This versatile new technique, which is being employed for a wide range of research applications (Felton and Brown, 1979; Weiner, 1979; Mohamed et al.,

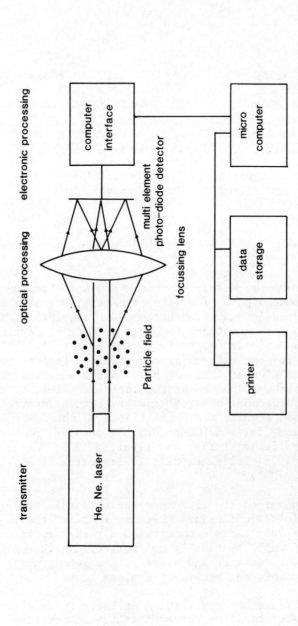

Figure 1. Schematic representation of laser diffraction equipment used for particle sizing.

1981) has a number of advantages which make it particularly suitable for studies of natural waters. The instrument is robust and readily adaptable to field operation requiring only a stabilised power supply. Analyses are automated and rapid. Most importantly, the instrument response is independent of the composition of the matrix supporting the sample and is therefore ideal for work in estuaries.

APPARATUS AND METHODS

Particle size distributions were measured using a Malvern Instruments Particle Sizer, Model 2200; this instrument is illustrated schematically in Figure 1. The light source incorporates a 2 mW HeNe laser and produces a parallel beam of monochromatic, coherent light. The detector consists of a Fourier transform lens which focuses light diffracted from the primary beam on to 30 semi-annular, concentric photo-diodes mounted normal to and centred on the beam axis. The radially symmetrical Fraunhofer diffraction pattern produced by particles within the beam is recorded as a light energy distribution across the detector rings. Microcomputer controlled scanning through the series of detectors at a rate of @ 25 sec^{-1} continuously updates the average light energy falling on each ring throughout the chosen measurement period. After the light energy distribution has been recorded the particle size distribution is calculated automatically or, if a fast sampling rate is required, the raw light energy data can be stored on magnetic tape or disk for processing later.

Particle size distributions are calculated using an heuristic approach. The computer makes an initial guess of the size distribution and calculates the diffraction pattern it would produce using Fraunhofer theory. The estimated light energy pattern is then iteratively refined to minimise the least-squares error coefficient between the observed and estimated light energy patterns, until adjustments become smaller than the resolving power of the instrument.

The magnitude of the final least-squares error provides a criterion for the goodness of the calculated fit. Size distributions are printed as a listing and/or a histogram of weight of material in each of 15 size bands as a percentage of the total weight in the measured range (1.2 to 118, 1.9 to 188 or 5.8 to 564 μm diameter according to the choice of focussing lens). Particle number and area distributions and other derived information can be computed using appropriate software and displayed or printed. Besides operating in this model independent fashion, which is essential for resolving multimodal size distributions, the light energy data can also be constrained to fit any of three monomodal particle size models: normal log-normal and Rosin Rammler, during the size distribution calculation. Each model

Table 1. Asymmetric frequency distribution models commonly used to describe naturally occurring particle populations.

Normal distribution

$$Nd = \frac{1}{N\sqrt{2\pi}} \exp \frac{(d - x)^2}{2N^2}$$

where Nd indicates the relative frequency of size d

Log-normal distribution

$$ld = \frac{1}{\ln(N)\sqrt{2\pi}} \exp \frac{(\ln(d) - \ln(x))^2}{2\ln(N)^2}$$

where ld is the weight frequency at size d

Rosin Rammler distribution

$$R = \exp(-d/x)^N$$

where R represents the normalised weight below size d

(see Table 1) employs two parameters to describe a unimodal distribution, x defining a principal size (the mean in the case of the normal distribution) and N a factor indicating the extent of spread about the principal size. The models differ in the extent of skewness they accommodate.

The normal distribution is rarely used for modelling distributions by weight as natural processes characteristically give rise to normal number distributions. However, it does have an application in very narrow distributions and has the advantage that the mean and standard deviation are familiar and easily manipulated concepts. Log normal distributions are typically generated by comminutive processes and have a wide application in industry. The Rosin Rammler model was originally devised to study the size distribution of crushed coal (Rosin and Rammler, 1933) but its application has been extended to include sand, ore and clay particle size distributions (Dapples, 1975). Aerosol spray droplet sizes tend to conform closely with the Rosin Rammler model. A fuller discussion of these models and their application to natural particle populations has been given by Lerman (1979).

During field measurements the instrument was electronically zeroed and a background signal was recorded using water from the estuary which had been filtered through a 0.45 µm pore-sized membrane filter and held in a 1 cm optical cell. The background signal, produced by stray light and optical defects, was automatically subtracted from each subsequent measurement and was regularly checked for consistency. Particle size measurements on estuarine samples were carried out, using the same cell, by integrating over 1000 detector scans (approximately 40 seconds) using the lens system appropriate to the 1.9 to 188 µm diameter range (see Figure 3 for a breakdown of the 15 size bands). Obscuration of the laser beam was checked during each measurement as an indicator of potential interference from multiple diffractions. When excessive obscuration occurred, a portion of the sample, filtered through a 0.45 µm membrane, was used to dilute the original sample.

Initially, we attempted to characterise particle size distributions in the estuary using a flow-through sample cell supplied by a submerged centrifugal pump; this procedure is used routinely for continuous recording underway of a number of aquatic constituents and properties (Morris et al., 1978, 1981). However, size distributions obtained in this way often bore little resemblance to those determined on discrete samples collected by bucket when the vessel was stationary. The discrepancies were found to be due partly to the formation of small (< 10 µm diameter) air bubbles within pumped samples. The primary problem with pumped samples however, was disaggregation: natural suspensions frequently contained loosely associated agglomerates. Samples collected by bucket however were not stable; rapid settling of larger particles was usually evident. These observations led us to adopt the following procedure, which is limited in its applicability to discrete sampling of surface water. Surface samples were abstracted from the estuary by dipping a bucket with minimum disturbance from the stationary vessel. The optical cell was immediately filled from the bucket and the diffraction pattern was recorded. A magnetic stirrer within the optical cell was used to maintain larger particles in suspension without seriously disrupting any aggregates. With careful application this method, although far from ideal, gave satisfactorily reproducible results both for replicate samples from the bucket and for successive observations on a single sample.

Two additional problems arising from the field deployment of the instrument required attention. Firstly, condensation on the optical cell faces caused serious interference when the sample temperature was appreciably below that of the ambient air. This was overcome by directing a stream of warmed air over the optical faces of the cell. Secondly, interferences arising from changes in ambient light intensity were overcome by encasing the optical system within an opaque cover.

Following the development of these operational procedures, the instrument was deployed in the Tamar Estuary on 16 July 1981. Particle size distributions of surface waters were recorded throughout the estuarine mixing profile as well as during repetitive traverses through the turbidity maximum zone. Basic environmental conditions (including salinity and turbidity) were recorded simultaneously.

RESULTS

Particle size distributions in the higher salinity (> 15°/oo) waters of the outer estuary were multimodal although there was a predominance of particles larger than 40 μm diameter. Occasional microscopic examinations showed that these larger particles were mainly planktonic. Particle size distributions in the freshwater input were also multimodal and dominated by large aggregates and particles. In contrast, samples collected from within the turbidity maximum zone showed much more regular particle size distributions which were essentially unimodal and asymmetric about a modal peak size which was appreciably smaller than those which dominated the fresh and more saline regions of the estuary. These general observations are illustrated by the data collected near the high water slack tide on 16 July 1981 shown in Figure 2. At this time a maximum turbidity of ca. 100 mg l^{-1} coincided with the freshwater-brackish water interphase some 4 km below the weir which marks the head of the estuary. The mean of the weight-size distribution, calculated from the best fit to a Rosin Rammler model ranged from greater than 100 μm diameter in the fresh water to 20-25 μm within the turbidity maximum zone. Further seaward, mean values increased to 40-60 μm diameter within the salinity range 1 to 10°/oo as the turbidity decreased to ca. 20 mg l^{-1}.

Residual least-squares errors showed that samples from within the turbidity maximum zone gave extremely good fits to a Rosin-Rammler model with no evidence of systematic errors. In fact, the Rosin Rammler size distributions closely matched those produced by the model independent routine for these samples. Normal and log-normal models were clearly inappropriate. The goodness of fit to the Rosin-Rammler model deteriorated progressively away from the core of high turbidity and the model became inadequate for all samples of turbidity less than 40 mg l^{-1}. For these samples, satisfactory size distributions could be obtained only by using the model independent fitting routine because of their multimodal nature.

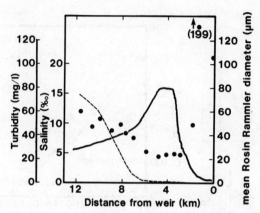

Figure 2. Variations in salinity (broken line), turbidity (full line) and mean Rosin-Rammler diameter of particle populations (dots) along the axis of the upper Tamar Estuary near slack high water on 16 July, 1981.

Variations in the particle-size distribution with changes in suspended load within the turbidity maximum zone recorded on 16 July 1981 are illustrated in Figure 3. Five samples were collected from the point of maximum turbidity between the time of strongest flood tide at 1511 hours and just after high water slack tide at 1904 hours. During this period the maximum turbidity decreased from greater than 400 mg l^{-1} to around 100 mg l^{-1} as tidally resuspended sediment settled. The size histograms obtained for these five samples, using the model independent routine are shown in Figure 3. The model independent and Rosin-Rammler fitting routines produced very similar shaped distributions and for comparison the modal values derived from the Rosin Rammler relationship have been superimposed on the model independent histograms in Figure 3. It is evident from this figure that as the concentration of suspended material in the turbidity maximum decreased towards slack water the particle size distribution became much broader and the modal diameter reduced from ca. 50 µm to a little more than 20 µm.

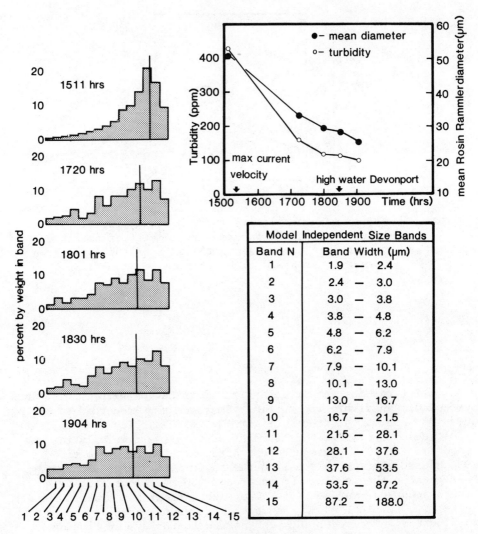

Figure 3. Histograms showing the model-independently computed size spectra of suspended particles collected sequentially from the core of the turbidity maximum in the Tamar Estuary on 16 July, 1981. The inset graph shows the decreases in turbidity and mean Rosin-Rammler diameter (which is also recorded on the histograms) accompanying the deposition of tidally resuspended sediment.

DISCUSSION

These results indicate that there are a number of different regimes within the surface waters of the estuary with respect to the size distribution of suspended particles. Except within the turbidity maximum zone, the particle size distributions were multimodal and dominated by relatively large particles. Particle size distributions within the surface waters of the turbidity maximum were not directly related to those of the marine or freshwater source material; the size distributions were essentially unimodal and asymmetric and larger particles were of less significance. A relatively sharp modal peak diameter was evident within the turbidity maximum when turbidities were greatest at peak tidal currents. This was attributed to tidal resuspension of sediment. The dominance of ca. 50 µm particles abated as the suspended load reduced with the approach of slack water. The suspended load within the turbidity maximum zone always remained appreciably higher than elsewhere in the estuary indicating the presence of a more or less permanently suspended population of particles trapped within this zone. This population was characterised by a broad spread of sizes about a relatively small modal peak diameter of 20-25 µm.

These observed changes in particle size composition can be correlated with changes in the chemical composition of suspended particles in the estuary. Loring et al., (1981) have shown that particles undergoing continuous tidally controlled cycling between sediment and water column within the turbidity maximum zone are enriched in quartz and impoverished in organic carbon relative to other suspended materials. Materials remaining trapped in permanent suspension within the turbidity maximum are richer in organic carbon, but are not as enriched as the marine and freshwater materials. Thus, both the particle size data and the chemical information support the hypothesis that particles within the turbidity maximum are actively selected. Basic descriptions of the processes involved have been given by Postma (1967) and by Schubel (1968) and analytical treatments have been given by Festa and Hansen (1978) and by Officer (1980).

The close conformity of the asymmetric, monomodal size distributions within the turbidity maximum zone to a Rosin Rammler distribution is surprising considering the selective processes that dominate the particle size distribution in this region. The underlying physical cause for the applicability of what is essentially a fragmentation model is not clear.

These preliminary results demonstrate the considerable potential offered by the laser diffraction technique for characterising particle size distributions in aquatic systems. Physically and chemically controlled particle aggregation and disaggregation within the estuary undoubtedly contribute in part to the observed size distributions. Large aggregates, by the nature of the attractive forces involved may only be loosely associated. The problems of sample integrity, which apply to all sizing techniques, therefore remain a major source of error which may never be overcome when discrete sampling techniques are applied. However, it appears technically feasible to develop a submersible laser diffraction instrument for observations in situ; such a development would allow the potential of this powerful technique to be fully realised.

ACKNOWLEDGEMENTS

This work forms part of the Estuarine Ecology Programme of the Institute for Marine Environmental Research, a component of the Natural Environment Research Council, and was partly supported by the Department of the Environment under Contract No. DGR 480/684.

REFERENCES

Burton, J.D. & Liss, P.S. (1976). Eds. Estuarine Chemistry. Academic Press, London. 229 pp.

Butler, E.I. & Tibbits, S. (1972). Chemical survey of the Tamar Estuary. I. Properties of the waters. Journal of the Marine Biological Association, U.K., 52, 681.

Dapples, E.C. (1975). Laws of distribution applied to sand sizes. Memoirs of the Geological Society of America, 142, 37.

Felton, P.C. & Brown, D.J. (1979). Measurement of crystal growth rates by laser diffraction. Paper presented at the 6th Annual Institute of Chemical Engineers Research Meeting. University College, London, 4-6 April, 1979.

Festa, J.F. & Hansen, D.V. (1978). Turbidity maxima in partially mixed estuaries: a two-dimensional numerical model. Estuarine and Coastal Marine Science, 7, 347.

Lerman, A. (1979). Geochemical processes, water and sediment environments. Wiley-Interscience, New York, 481 pp.

Loring, D.H., Rantala, R.T.T., Morris, A.W., Bale, A.J. & Howland, R.J.M. (1981). The chemical composition of suspended particles in an estuarine turbidity maximum zone. Paper presented at the Symposium on the Dynamics of Turbid Coastal Environments. Halifax, Canada. September, 1981.

Mohamed, N., Fry, R.C. & Wetzel, D.L. (1981). Laser Fraunhofer diffraction studies of aerosol droplet size in atomic spectrochemical analysis. Analytical Chemistry, 53, 639.

Morris, A.W. (1978). Chemical processes in estuaries: the importance of pH and its variability. In: Environmental Biogeochemistry and Geomicrobiology. Vol. 1, the Aquatic

Environment (Krumbein, W.E., ed.). Ann Arbor Science, Ann Arbor.

Morris, A.W., Mantoura, R.F.C., Bale, A.J. & Howland, R.J.M. (1978). Very low salinity regions of estuaries: important sites for chemical and biological reactions. Nature, London, 274, 678.

Morris, A.W., Bale, A.J. & Howland, R.J.M. (1982a). The Dynamics of estuarine manganese cycling. Estuarine, Coastal and Shelf Science, 14, 175.

Morris, A.W., Bale, A.J. & Howland, R.J.M. (1982b). Chemical variability in the Tamar Estuary, southwest England. Estuarine, Coastal and Shelf Science, 14, 649

Morris, A.W., Loring, D.H., Bale, A.J., Howland, R.J.M., Mantoura, R.F.C. & Woodward, E.M.S. (1982c). Particle dynamics, particulate carbon and the oxygen minimum in an estuary. Oceanologica Acta, (In press).

Officer, C.B. (1980). Discussion of the turbidity maximum in partially mixed estuaries. Estuarine and Coastal Marine Science, 10, 239.

Postma, H. (1967). Sediment transport and sedimentation in the estuarine environment. In: Estuaries (Lauff, G.H., ed.). American Association for the Advancement of Science, Washington.

Readman, J.W., Mantoura, R.F.C., Rhead, M.M. & Brown, L. (1982). Aquatic distribution and heterotrophic degradation of polycyclic aromatic hydrocarbons (PAH) in the Tamar Estuary. Estuarine, Coastal and Shelf Science, 14, 369.

Rosin, P. & Rammler, E. (1933). The laws governing the fineness of coal. Journal of the Institute of Fuel, 7, 29.

Schubel, J.R. (1969). Size distributions of the suspended particles of the Chesapeake Bay turbidity maximum. Netherlands Journal of Sea Research, 4, 283.

Weiner, B.B. (1979). Particle and Spray Sizing using laser diffraction. SPIE. Optics in quality Assurance II, 170, 53.

DETECTING COMPOSITIONAL VARIATIONS IN FINE-GRAINED SEDIMENTS BY TRANSMISSION ELECTRON MICROANALYSIS

P.J. Kershaw

Ministry of Agriculture, Fisheries and Food
Directorate of Fisheries Research
Fisheries Laboratory
Lowestoft, Suffolk UK

INTRODUCTION

Clay material, by definition, is composed of very small particles which are difficult to separate out individually. This initially delayed the identification of the mineral constituents of soils and fine-grained sediments although 'the ancients' were able to identify varieties of 'earths' by the properties they exhibited (Theophrastus, ca 300 BC, in Mackenzie, 1975) and subsequently these differences were found to be expressions of mineralogical variations.

The identification of clay mineral groups became possible with the introduction of X-ray diffraction techniques in the 1920s and 1930s and this remains the most widely used method. Because of the complexities of clay mineral identification and structural interpretation XRD data are supplemented frequently by some other form of analysis, typically differential thermal analysis or infrared spectrometry. Such techniques analyse the 'total' sample, and when this consists of a mixture of different minerals, interpretation of the resulting spectra may present difficulties.

Developments in the field of electron microscopy now permit rapid mineral identification and quantitative chemical analysis of individual clay-sized particles and groups of particles. It is not intended that such methods should replace XRD as the primary tool in clay mineral analysis but rather that they should be used to provide additional information which may not be available by other common forms of analysis.

© Crown Copyright

MINERAL IDENTIFICATION AND ANALYSIS BY TEM

Quantitative Microanalysis

The microanalysis system used in this investigation consisted of an AEI 100 kV transmission electron microscope (EM6G) fitted with an energy dispersive Si(Li) Kevex detector and a Link Systems model 290 analyser. The 200 μA electron beam had a minimum diameter of 6 μm. The relatively large beam diameter limited single-grain analysis to particles greater than 0.2 μm in diameter. This was partly a consequence of the instrument's age and partly due to modification of the electron optics to allow the incorporation of the microanalysis system (Fitzgerald and Storey, 1979). On modern equipment minimum beam diameters of 100 Å have been reported (Beaman and File, 1976).

All matter will emit X-rays when struck by electrons, or X-ray photons, of sufficient energy. As the atomic number of the element increases the energy of the emitted X-ray photon is proportionately higher and characteristic of that element. An energy dispersive detector allows the whole spectrum of characteristic radiation to be recorded in a single measurement.

When X-rays enter the detector crystal they are absorbed giving rise to electrons which dissipate their energy within the medium by ionising the silicon to form mobile charge carriers. The number of electron-hole pairs produced is proportional to the energy of the photon and follows a near-Gaussian distribution. The relatively poor resolution of energy-dispersive detectors can lead to inter-element interference in the form of overlap of closely-spaced peaks. This becomes significant at the low energy end of the spectrum particularly for the elements sodium (1.08 KeV), magnesium (1.26 KeV), aluminium (1.48 KeV), and silicon (1.76 KeV). If corrections are not applied then data are limited to semi-quantitative interpretation although it is still possible, using characteristic element ratios, to identify phases of differing composition. To overcome this inaccuracy an iterative peak-stripping computer program was employed which allowed the simultaneous correction of up to three peaks for overlap and background contribution.

Quantitative analysis of individual mineral particles became possible using the technique of thin-specimen microanalysis described by Cliff and Lorimer (1975). If the specimen thickness is less than 1000-1500 Å the differential generation and absorption of X-ray photons within the specimen is insignificant and the intensity of the characteristic line becomes proportional to the concentration of that element in the specimen. Concentrations are calculated, ratioing for sets of element pairs, using the expression:

$$\frac{C_A}{C_B} = \frac{I_A}{I_B} k \qquad (1)$$

where

$$k = \frac{g_A \, \varepsilon_A}{g_B \, \varepsilon_B} \qquad (2)$$

$C_{A,B}$ = concentration of elements A and B in specimen;

$I_{A,B}$ = intensities of the characteristic lines;

$g_{A,B}$ = generation efficiency;

$\varepsilon_{A,B}$ = detector efficiency.

Cliff and Lorimer (1975) have described a method to allow the calculation of the factor k using materials of known composition (McGill and Hubbard, 1981).

Sample Preparation

Clay mineral reference materials were wet ground in distilled water with an agate pestle and mortar, transferred to test tubes, suspended in distilled water, diluted until almost colourless, and placed in an ultrasonic bath for 30 minutes. Specimens were prepared by applying a single drop of the dilute suspension to a carbon film supported on a copper or carbon-coated nylon microgrid (2 mm diameter). The use of nylon grids is preferable as this eliminates possible interference by the copper L line. These procedures were found to provide a uniform distribution of grains on the specimen grid at a density giving adequate numbers for analysis without causing inter-grain interference. Only an extremely small proportion of the sample is microanalysed and it is therefore important to avoid contamination during specimen preparation.

Identification of Clay Mineral Particles

A set of sixteen reference minerals provided representatives from the main clay mineral groups (Table 1). The composition of each was characterised by a minimum of six single-grain analyses (Table 2). The data were presented as element:silicon ratios. To allow the mineral identification of individual particles from fine-grained sediment samples a method was developed which was easy to prepare and which provided fairly rapid mineral identification.

Table 1. Source and Coding of Reference Minerals Used in the Electron Microanalysis of Clay Minerals

Mineral Group	Mineral Species	Locality	Source	Analysis Code	Number of Analyses
Kaolin	Kaolinite	-	MI	MIK	6
	Kaolinite	CMS 5	IGS	IGSK	8
	Dickite	CMS 14	IGS	IGD	6
Mica	Muscovite	-	DU	MUG	8
	Biotite	-	DU	BTG	10
	Illite (Fithian)	Illinois	MI	MIL, FIG	8, 21
Smectite	Wilkinite	-	MI	MIW	10
	Nontronite	-	MI	MIN	6
	Nontronite	CMS 33B	IGS	IGN	6
	Saponite	Spain	IGS	IGS	6
	Na-mont.	CMS 27	IGS	NM	6
	Ca-mont.	Baulking	IGS	CMB	5
Vermiculite	Vermiculite	Kenya	IGS	IGV	6
Chlorite	Prochlorite	-	DU	CHG	9
Mixed Layer	K-bentonite	Woodbury	IGS	IGBW	6
	Illite-smectite	Firth of Forth	MX	MOG	7

Source: MI   Macaulay Institute of Soil Research, Aberdeen
IGS   Institute of Geological Sciences, London
DU   Geology Department, University of Dundee
MX   Mineral extraction (Kershaw, 1980).

'Templates' of the sixteen reference minerals were produced, representing the range of the element:silicon ratios encountered. The element ratios of unknown minerals were plotted on overlays allowing direct comparisons to be made (Fig. 1). This method of 'pattern-fitting' has advantages over a more rigid system, particularly if the mineralogy of the unknowns does not exactly match that of any of the reference materials, in much the same way as the more sophisticated computer matching approach of Rubin and Maggiore (1974).

Raw peak intensities of unknowns were readily corrected and converted into normalised oxide weight percentages (Table 3) allowing the calculation of the minerals' structural formulae. An example of the derivation of a structural formula, from a potassium-bentonite analysis, is given in Table 4. The chemical composition is presented as oxide weights normalised to 100% ignoring water content. The number of moles of each oxide is found by dividing the oxide weight percent (column 1) by its molecular weight (2). This value is multiplied by the number of cations in each mole to give the atomic proportion of cations (3) and by the number of oxygens to give the atomic proportion of oxygen (4). The total proportion of oxygen (summation of the proportion for each oxide) is equivalent to

Table 2. Element X-ray Intensity Ratios of Sixteen Reference Minerals

Mineral		Al:Si	K:Si	Fe:Si	Mg:Si	Ca:Si
Kaolinite MIK						
	Mean	0.75	0.06	0.02		
	Range	0.68-0.82	0.01-0.23	0.001-0.05		
Kaolinite IGSK						
	Mean	0.82	0.02	0.01		
	Range	0.77-0.86	0.01-0.04	0.003-0.01		
Dickite IGD						
	Mean	0.52	0.02	0.03		0.002
	Range	0.43-0.72	0.001-0.05	0.01-0.05		0.001-0.006
Muscovite MUG						
	Mean	0.64	0.29	0.05		
	Range	0.55-0.69	0.20-0.32	0.003-0.09		
Biotite BTG						
	Mean	0.44	0.30	0.93	0.03	
	Range	0.40-0.55	0.16-0.38	0.63-1.56	0.00-0.05	
Illite MIL						
	Mean	0.51	0.21	0.11		
	Range	0.41-0.65	0.17-0.27	0.06-0.14		
Wilkinite MIW						
	Mean	0.31		0.07		0.07
	Range	0.24-0.36		0.06-0.09		0.1-0.17
Nontronite MIN						
	Mean	0.06		0.83		0.06
	Range	0.04-0.07		0.77-0.87		0.04-0.07
Nontronite IGN						
	Mean	0.15	0.004	0.66	0.004	0.04
	Range	0.13-0.18	0.001-0.01	0.59-0.72	0.002-0.007	0.03-0.05
Saponite IGS						
	Mean	0.03	0.03	0.03	0.31	0.03
	Range	0.01-0.05	0.01-0.06	0.002-0.05	0.27-0.35	0.004-0.04
Na-mont. NM						
	Mean	0.33	0.01	0.07	0.02	0.01
	Range	0.32-0.35	0.01-0.01	0.06-0.07	0.01-0.02	0.01-0.01
Ca-mont. CMB						
	Mean	0.34	0.004	0.07	0.03	0.01
	Range	0.34-0.35	0.004-0.007	0.07-0.08	0.02-0.03	0.01-0.02
Vermiculite IGV						
	Mean	0.36	0.02	0.21	0.51	0.03
	Range	0.33-0.42	0.01-0.02	0.19-0.23	0.48-0.62	0.02-0.05
Prochlorite CHG						
	Mean	0.41	0.01	0.43	0.71	0.02
	Range	0.38-0.45	0.01-0.03	0.40-0.45	0.65-0.77	0.01-0.04
K-bentonite IGBW						
	Mean	0.43	0.17	0.04	0.06	0.03
	Range	0.42-0.43	0.17-0.18	0.04-0.04	0.05-0.07	0.03-0.03
Illite-smectite MOG						
	Mean	0.41	0.11	0.24	0.10	
	Range	0.39-0.43	0.10-0.12	0.22-0.26	0.09-0.10	

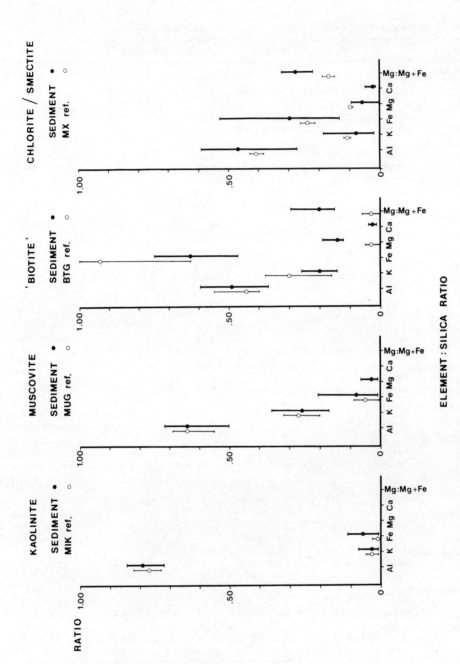

Fig. 1. Clay mineral phases identified in the sediment by comparison with four reference minerals using transmission electron microanalysis.

Table 3. Conversion of Element Peak Intensity to Oxide Weight Percentage

	Na	Mg	Al	Si	K	Ca	Fe
Peak Counts	n.d.	2707	20290	47046	8135	1353	1978
k-value	3.00	1.70	1.15	1.00	1.08	1.06	1.37
Corrected Integral		4602	23334	47046	8786	1434	2710
Metallic Fraction in Oxide	0.742	0.603	0.529	0.467	0.830	0.715	0.699

	Na_2O	MgO	Al_2O_3	SiO_2	K_2O	CaO	Fe_2O_3
Corrected Oxide Integral		7632	44109	100741	10585	2006	3877
Normalised to 100%		4.517	26.108	59.628	6.265	1.187	2.295

Table 4. Calculation of Structural Formulae - Example: K-bentonite

Oxide	1	2	3	4	5	6	
	Weight (%)	Molecular weight	Proportion of Cation	Proportion of Oxygen	Number of ions	Allocation	
SiO_2	59.628	60.085	0.9924	1.9848	7.2876	7.29	
							8.00 Tetrahedral
						0.71	
Al_2O_3	26.108	101.961	0.5121	0.7682	3.7606		
						3.05	
Fe_2O_3	2.295	159.692	0.0287	0.0431	0.2108	0.21	4.08 Octahedral
MgO	4.517	40.304	0.1121	0.1121	0.8232	0.82	
K_2O	6.265	94.203	0.1330	0.0665	0.9767	0.98	
							1.14 Interlayer
CaO	1.187	56.079	0.0212	0.0212	0.1557	0.16	
Na_2O	n.d.	61.979					

Total Proportion of Oxygen = 2.9959

Multiplying Factor 'n' = $\frac{22}{2.9959}$ = 7.3435

Structural Formula $K_{0.98}Ca_{0.16}(Si_{7.29}Al_{0.71})(Al_{3.05}Fe_{0.21}Mg_{0.82})O_{20}(OH)_4$

the number of oxygens per formula unit. By dividing this number by the total proportion of oxygen a multiplying factor ('n') is obtained. Molecular proportions, or number of moles, are converted to number of cations per structural unit by multiplying the atomic cation proportions by 'n' (5).

Ions were assigned to positions in the lattice following the method of Nagelschmidt (1938). All the silicon and a proportion of the aluminium was allocated to fill the tetrahedral sites. The remaining aluminium, together with iron and magnesium, was assigned to the octahedral layer. Potassium, calcium and sodium were assumed to occupy interlayer positions and to act as charge compensators. The 'ideal' crystal lattice of the 2:1 group of clay minerals (micas and smectites) was assumed to contain 20 oxygens and 4 hydroxyl ions ($220, 2H_2O$) per formula unit and of the 1:1 group of minerals (kaolins) 10 oxygens and 8 hydroxyl ions ($140, 4H_2O$).

Microanalysis of a Cohesive Sediment

The thin-specimen techniques described were applied to a sample of cohesive bottom sediment from the Forth Estuary. Prior to analysis, this sample had been subjected to a number of treatments designed to promote maximum disaggregation and dispersion of the sediment into individual grains and to remove surface coating effects from the subsequent chemical analyses: viz. washing in distilled water to remove excess soluble salts; dispersal in a 0.2 g l^{-1} solution of sodium polymetaphosphate (83%) and sodium carbonate (17%) to inhibit flocculation; rigorous hydrogen peroxide (30%) treatment to remove the bulk of the organic matter; and sodium citrate-dithionite treatment to remove iron coatings. A detailed account of the procedures adopted and their possible consequences has been described elsewhere (Kershaw, 1980). None of the treatments were considered to cause appreciable compositional change to the minerals analysed.

Once the pretreatments were completed the sample was size fractionated by a combination of gravity settling and centrifugation into 7 size fractions: < 0.2, 0.2-0.5, 0.5-1.0, 1-2, 2-6, 6-12 and > 12 µm e.s.d. (equivalent spherical diameter). Material below 12 µm was of such a size as to make further reduction in size, by grinding, both unnecessary and undesirable. The fraction over 12 µm was reduced to a fine powder by grinding in an agate mortar. Specimen grids of each of the size fractions were prepared and analysed by TEM. In addition, the mineralogy of this material was established by X-ray diffraction analysis (see Kershaw, 1980).

The analyses were grouped, using the template-matching method, into four clay mineral phases. Most of the grains were clearly identified as belonging to either the kaolinite group or the muscovite group of minerals. The identity of the remaining grains was more ambiguous but it was possible to identify a potassium-, iron-, and magnesium-rich 'biotite' phase. The remaining grains were grouped together as a chlorite-smectite phase (Table 5).

Table 5. X-ray Intensity Ratios of Minerals Identified in Sample 262/3

Mineral		Al:Si	K:Si	Fe:Si	Mg:Si	Na:Si
Kaolinite						
	Mean	0.77	0.02	0.05		
	Range	0.68-0.84	0.00-0.07	0.01-0.11		
Muscovite						
	Mean	0.61	0.26	0.09	0.02	
	Range	0.43-0.67	0.16-0.35	0.03-0.20	0.01-0.05	
'Biotite'						
	Mean	0.47	0.20	0.62	0.14	
	Range	0.37-0.59	0.14-0.26	0.47-0.74	0.12-0.18	
Chlorite/Smectite						
	Mean	0.47	0.08	0.30	0.08	0.02
	Range	0.28-0.59	0.02-0.18	0.14-0.53	0.00-0.09	0.01-0.04

In addition to the quantitative analysis of thin grains a number of qualitative analyses were made of thick or opaque grains which formed the bulk of the sample in the > 1 μm size range. Iron particles (iron peak only), calcium carbonate microliths (calcium only) and feldspar grains (calcium-rich) were identified in the coarser fractions. Small (relative to the surrounding clay flakes), equi-dimensional grains of quartz (silicon only) were common throughout the size range. Biogenic debris, in the form of diatom tests (silicon only) was observed in the 0.2 to 6.0 μm size range. The spectra of many opaque grains showed similarities with the spectra of the 'biotite' and 'chlorite/smectite' phases and it seems probable that the abundance of thin grains of kaolinite and muscovite reflected the ease with which these minerals could cleave without significantly reducing their overall dimensions - a consequence of their higher degree of crystallinity.

Estimation of Bulk Composition by Multi-grain Analysis

The single-grain method did not permit the modal analysis of the complete specimen because only grains which were sufficiently thin could be determined. The analysis would thus favour those minerals which preferentially yielded thin flakes. It would be possible to carry out a modal analysis of thick particles if the compositions of the phases present were sufficiently well defined. Obtaining a sufficient number of determinations would be extremely time-consuming. As the particle-size decreased the number of thick and opaque grains rapidly decreased but it became increasingly difficult to find isolated grains which yielded sufficient counts to give reliable analyses. This was a consequence of the relatively large electron beam diameter (minimum diameter 6 μm). Single-grain analyses were possible on all but the smallest size fraction (< 0.2 μm diameter) and this established the number of phases present and their composition throughout the size range. In the four finest size fractions (< 2 μm diameter) most of the grains were sufficiently thin to permit their quantitative analysis and they often formed a near-continuous single-grain-thick layer. By defocussing the beam to about 24 μm diameter, it was possible to analyse up to 800 grains simultaneously and to obtain the bulk chemical composition of each size-fraction by averaging the results of six separate analyses.

From a knowledge of the bulk chemistry and the composition of the phases present in the finer-grained fractions it was possible to determine the proportion of each phase in the specimen by solving a series of simultaneous equations of the form:

$$A_u U + A_w W + A_x X + A_y Y + A_z Z = A_T$$

where U, W, X, Y, Z represent the proportion of each mineral phase present and A represents the proportion of a particular element in each phase (A_u - A_z) and in the bulk analysis (A_T). Because there were five phases, four clay phases and silica, at least five equations were required. The most precisely determined elements were used: magnesium, aluminium, silicon, potassium and iron, normalised to 100%. Some sodium and calcium values were unreliable due to interference from copper L and potassium Kβ lines respectively. Copper L line interference can be eliminated by the use of nylon specimen grids. The contribution of potassium Kβ to the calcium Kα peak can be corrected for by the use of the potassium Kβ/Kα quotient which approximates to 0.12 (Slivinsky and Ebert, 1972). The matrix was solved utilising a NAG routine (F04 AUA). A similar method was successfully applied to the analysis of slate dusts (Pooley, 1977), and to the analysis of marine metalliferous sediments (Dymond and Eklund, 1978). In both cases the phases present could be clearly differentiated on the basis of their chemistry.

It may be more difficult to separate clay mineral mixtures into distinct phases based on chemical differences without additional structural information. Pearson (1978) used published values of clay mineral phase composition to obtain a quantitative clay mineral analysis of mudrocks from whole-rock X-ray fluorescence data.

The proportions of the five phases in each size fraction were normalised to 100% (Fig. 2). The most striking trend observed was the rapid increase in 'biotite', an iron-rich and potassium-rich phase, as the grain-size decreased. This was accompanied by a decrease in the proportion of kaolinite. The percentage of muscovite and silica fluctuated, showing no clear trend, and the proportion of the 'chlorite/smectite' phase remained constant.

Bulk chemical analysis established the presence of significant amounts of free silica, throughout the size range, in the form of detrital quartz grains, cryptocrystalline coatings and biogenic debris. This was not clearly shown by X-ray diffraction analysis, partly because of the coincidence of the 003 illite and 101 quartz peaks and partly because of a reduction in intensity in the finer-grained fractions caused by peak broadening and the presence of amorphous silica.

The identification of a 'chlorite/smectite' phase suggested the presence of an iron-rich chlorite mineral. This was later confirmed by detailed analysis of the X-ray diffraction pattern. The relatively high potassium levels implied a degree of interlayering with illite. A similar template-pattern was produced by the mixed-layer reference mineral potassium-bentonite. It is possible that the smectite component formed a continuous gradation into the 'biotite' phase.

There was a rapid increase in the proportion of the 'biotite' phase as grain-size decreased. It is suggested that this represented an authigenic transformation approaching glauconite in composition rather than unaltered, detrital biotite. There was supporting evidence from X-ray diffraction analysis to indicate the presence of illite-smectite interlaying in the finest fractions. Glauconite is an iron-rich, mixed-layer illite-smectite mineral (Thompson and Howe, 1975) formed in a mildly reducing environment, especially in sediments dominated by fine sand- to coarse clay-sized particles, periodically disturbed by marine bottom currents. Such were the conditions existing at the sampling sites.

The chemical identification of kaolin in every size fraction served as confirmation that the 7 Å mineral identified by X-ray diffraction analysis was a dioctahedral kaolin mineral. The

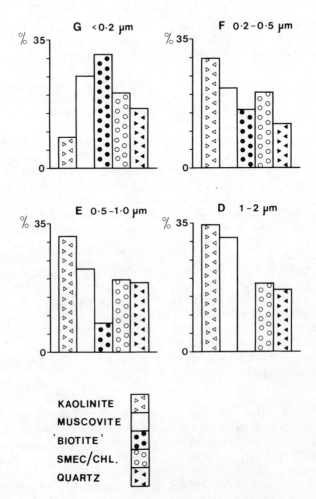

Fig. 2. Mineralogical composition of the four finest grain-size fractions (G to D, < 0.2 to 2 μm respectively) from transmission electron microanalysis.

euhedral, hexagonal appearance of many of the grains implied that the mineral was kaolinite rather than halloysite which tends to have a tubular form. Bulk chemical analysis suggested that the 10 Å diffraction peak was largely contributed by a muscovite-type mica.

Compositional Variations in Muscovite/Illite Grains

A number of compositional variations were observed in analyses of muscovite and illite grains which appeared to be associated with grain size. An apparent dependence of the potassium content of muscovite on particle size was established in the size fractionated material of the cohesive sediment sample. The aluminium, iron and magnesium ratios of this phase showed an even distribution throughout the size range of analysed grains. However, the potassium: silicon ratio was strongly polarised around a grain size of approximately 1 µm: higher values tending to occur in the coarser fractions and lower values in the finer fractions (Fig. 3). A Chi square (χ^2) test was run to test the dependence of the potassium content (K:Si ratio $\gtrless$ 0.26) on the particle size (1 µm diameter). The potassium content and the grain size were dependent on each other at the 5% level of significance (χ^2 = 9.16). The aluminium, iron and magnesium ratios, and hence the overall chemical composition, were independent of particle size (χ^2 = 0.24, 2.03 and 0.33 respectively). The reduction in potassium content in the smaller diameter grains was therefore a direct consequence of a decrease in grain size rather than a difference in the composition of the unit layers.

Preferential loss of low atomic number elements may occur under electron bombardment (Read, 1975; McGill, 1980). This effect is more pronounced when using a narrow-focussed, high intensity beam. Pooley (1977) noted a 10% loss of magnesium from crocidolite fibres which exhibited obvious beam damage. Beaman and File (1976) reported an apparent compositional variation of chrysotile fibres with grain size. They concluded that magnesium was lost from particles of less than 0.4 µm diameter. As the beam was adjusted according to the size of the fibre smaller particles were subject to a higher intensity beam. Using thin specimens less of the electron beam is absorbed so the problem of overheating and subsequent mobility is less critical. To minimise preferential loss of potassium from small illite grains a fixed counting time of 200 s real time and a constant beam diameter, and hence electron intensity, were maintained for all quantitative analyses. However, if smaller particles possess a higher degree of crystal imperfection some preferential loss of lower atomic number elements may be unavoidable. This tendency will be opposed by a corresponding increase in potassium-binding energy as discussed below.

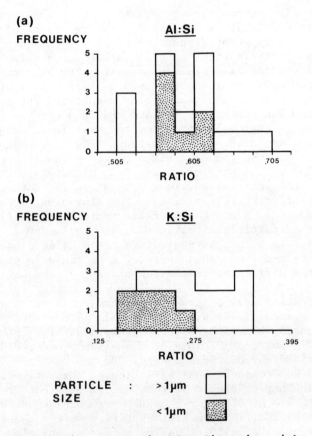

Fig. 3. Distribution of Al:Si and K:Si ratios with respect to grain size from analyses of muscovite grains in the cohesive sediment sample.

COMPOSITIONAL VARIATIONS IN FINE-GRAINED SEDIMENTS 101

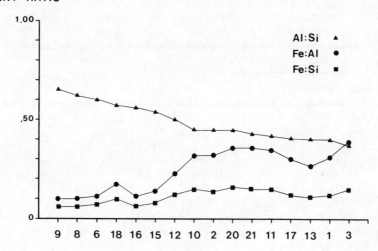

Fig. 4. Variations in element ratios demonstrating the progressive replacement of aluminium by iron and silicon from analyses of illite grains (FI series), ranked in order of decreasing Al:Si ratio.

Investigations into potassium behaviour in illites were continued using the reference mineral Fithian Illite. Because the material was fairly homogenous and lacked impurities adequate counting rates were maintained by analysing a suitable number of equal-sized particles simultaneously. Thus, for the largest particle diameter, a single grain filled the beam area and progressively more grains were analysed as the particle size decreased until, at the smallest diameters, the analysed material consisted of near-continuous mono-layers of very fine crystals. Grain thickness, as estimated from the intensity of the transmitted image appeared to be independent of the overall plate size. This was consistent with the observations of Reichenbach and Rich (1975) that small mica particles split by cleavages normal, rather than parallel, to the unit layer, resulting in grains of similar plate thickness but differing areas.

To avoid introducing bias the FI analyses were ranked in order of grain size before any of the chemical data were processed. These analyses revealed a series of distinctive trends when ranked in order of decreasing grain size (Fig. 4). The lowering of the Al:Si ratio reflected a progressive decrease in the amount of tetrahedral replacement of silicon (Si^{4+}) by aluminium (Al^{3+}) accompanied by a progressive increase in the octahedral replacement of aluminium by iron (Fe^{3+}). The substitution of tetrahedral Si^{4+} by Al^{3+} ions produces a charge deficit. This net negative layer charge is increased if there are insufficient cations to fill octahedral vacancies. It is compensated by the accumulation of K^+ ions, or other suitable cations (Na^+, Mg^{2+}, Ca^{2+}, NH^{3+}), in interlayer positions. To ensure that the trend in the Al:Si ratio was not being caused by a systematic error due to poor counting statistics or element interference, the count rate was monitored. Apart from the detector efficiency, which remains constant, the count rate is a function of (i) particle thickness; (ii) areal extent of material under the beam; (iii) differing ability of elements to produce photons; (iv) electron beam density; (v) mass absorption, which depends on the composition and thickness of the specimen. In the context of this study: (iii) and (iv) were constant, (v) could be neglected and (i) appeared to vary little. The main factor affecting the count rate was (ii), the amount of material under the beam. The trend in the Al:Si ratio was independent of the count rate (Fig. 5) implying that it was not an artefact of the method but due to variations in the chemical composition of the material.

Changes in the Al:Si ratio with diminishing grain-size have been reported elsewhere. Mackenzie and Milne (1953), investigating the effect of dry grinding on muscovite grains, observed that there was an increase in the Al:Si ratio with diminution of grain-size. They interpreted this as a loss of Si^{4+} ions due to the exposure of tetrahedra by rupture normal to the unit layers. Reichenbach and Rich (1975) concluded that the effect of artificially reducing particle size could be to increase or decrease the Al:Si ratio depending on the mechanical device employed. Filing and wet grinding are generally considered to be less disruptive to the crystal lattice.

There was a strong correlation between the weight percentages of Al_2O_3 and Fe_2O_3 ($r = -0.8879$) and an overall increase in the weight percentage of silica with decreasing grain size (Fig. 6). The proportion of magnesium showed a gradual increase implying that some of the octahedral vacancies were filled by Mg^{2+}. The amount of potassium decreased through the large and medium grain-sizes but levelled off below a particle diameter of about 1.0 μm. As particle size decreases more interlayer surfaces become exposed and K^+ ions may be exchanged more easily. However, given the finite plate thickness hypothesis, this loss may be expected to level off once rupture of the plate becomes normal, rather than parallel, to the unit layer. Although the proportion of potassium was lower in the smaller grains the amount of tetrahedral substitution, which controls the fixation of compensator ions, was correspondingly reduced (Table 6).

COMPOSITIONAL VARIATIONS IN FINE-GRAINED SEDIMENTS

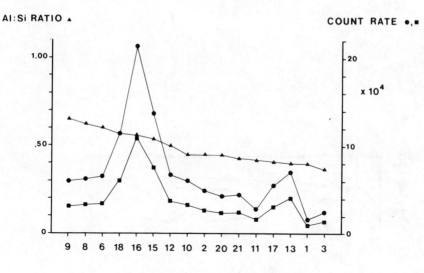

Fig. 5. Relationship of the Al:Si ratio (▲) to the total count rate (●) and silicon-only count rate (■) from analyses of illite grains (FI series).

Table 6. Charge Compensation in Illite Grains (MIL Series; Arranged in Order of Decreasing Al:Si Ratio)

MIL No.	Al:Si Ratio	Charge Deficit		Net Layer Charge, Negative	K^+ Interlayer Ions, Positive	Overall Charge
		Tetrahedral	Octahedral			
4	0.65	1.52	0.60	2.12	1.31	−0.81
1	0.64	1.57	0.66	2.23	1.36	−0.87
6	0.53	1.14	0.84	1.98	1.21	−0.77
8	0.52	1.09	0.78	1.87	1.16	−0.71
3	0.47	0.82	0.69	1.51	1.01	−0.50
5	0.47	0.86	0.69	1.55	0.93	−0.62
2	0.41	0.60	1.05	1.65	0.99	−0.66
7	0.41	0.66	1.08	1.74	1.05	−0.69

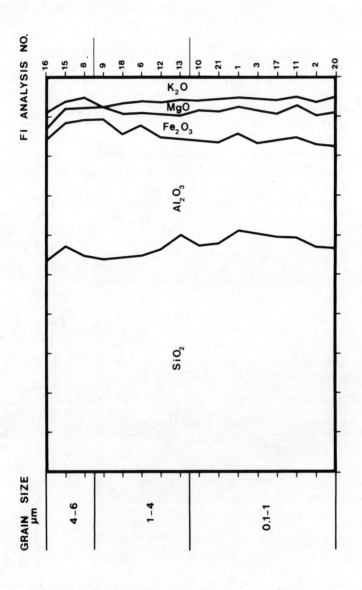

Fig. 6. Variation of chemical composition with particle size from analyses of illite grains (FI series), normalised to 100% (divisions on abscissa 10%).

Explanations of potassium behaviour in fine-grained micas have largely been based on ion-exchange studies. It has been found consistently that the removal of potassium becomes increasingly difficult as grain-size diminishes. Norrish (1972) suggested a mechanism to explain these observations. Because interlayer K^+ ions occur in 'holes' in the tetrahedral layer they lie adjacent to the octahedral hydroxyl groups. The bonding of interlayer potassium is very dependent on the orientation of the OH bond within the unit layer (Giese, 1971). The removal of one K^+ ion and its replacement by an hydrated Na^+ ion causes the OH bond angle (the angle it makes with the a-b crystallographic plane) to increase. This has the effect of decreasing the bond angle of the adjacent OH group which causes the remaining K^+ ion to be bonded more strongly. This mechanism will result in the formation of interlayer regions depleted in potassium adjacent to regions in which it is held more strongly leading to the formation of regularly interstratified mica-smectite/vermiculite structures.

POTENTIAL SOURCES OF CLAY MINERALS IN THE FORTH ESTUARY

The source of the clay minerals identified in Forth Estuary bottom sediments may be readily attributed to the varied solid and drift geology of the Central Region. Most of the River Forth catchment area is underlain by Carboniferous sediments and Devonian sediments and lavas. The clay mineralogy of the Carboniferous sediments is dominated by illite and kaolinite. Kaolinite is dominant in the Upper Carboniferous and, in sandstones, is always in a highly crystalline, morphologically-perfect form. The Lower Old Red Sandstone marls contain abundant dioctahedral mica and interstratified smectite-chlorite and include mica, chlorite, montmorillonite and mica-smectite. The lavas contain abundant trioctahedral smectite, saponite, kaolinite, illite, and a variety of dioctahedral interstratified minerals. Much of the area is covered by glacially-derived material which had its origins in the Grampian Highlands region. Such material provides an abundant source of micas, smectite, and chlorite minerals (Wilson, 1971; Wilson et al., 1972).

DISCUSSION

An energy-dispersive microanalytical system has a number of advantages and limitations. It is relatively inexpensive, easy to operate and accurate, and produces analyses at a faster rate than all but the most fully automated crystal spectrometers. Combined with a transmission electron microscope it allows the simultaneous recording of a specimen's chemical composition and its electron diffraction pattern. The thin-specimen method yields rapid quantitative analyses and not only aids mineral identification but permits the observation of intergrain variations which may be undetected in bulk analysis.

The major limitation to the methods followed in this study was the relative paucity of sufficient numbers of thin grains yielding high count rates. This problem could be partially solved by using a smaller beam diameter whilst being aware that specimen damage and loss or migration of volatile elements may become critical and require rigorous monitoring and standardisation of operating conditions.

The chemical data obtained by microanalysis are a fundamental reflection of the chemical composition of the sample because the production of characteristic X-rays within the specimen is independent of the nature of the material or its chemical state. In contrast, the shape and intensity of characteristic X-ray diffraction peaks are very dependent on the chemical and physical state of the material and its crystallite size, and may be influenced by variations in the procedure of specimen preparation and presentation to the X-ray beam.

Both X-ray diffraction analysis and transmission electron micro-analysis suffer disadvantages which may limit mineral identification but these can largely be overcome by combining the chemical data with the measurements of the materials' physical properties. In addition, the combination of electron-optical and analytical techniques may reveal the presence of constituents that are not revealed by X-ray diffraction either because they are present in too small quantities or because they are amorphous to X-rays.

SUMMARY

The chemical analysis of various clay minerals was undertaken using a transmission electron microscope equipped with an energy-dispersive detector. The identification of the principal phases present in the size-fractionated material of a cohesive bottom sediment sample was achieved by matching element:silicon ratios of 'unknowns' with templates of element ratios obtained from 16 reference minerals.

Bulk analysis of this material was made possible by de-focussing the electron beam to allow the simultaneous analysis of several hundred thin grains. By applying the knowledge of the phase compositions to the bulk chemical analysis it was possible to estimate the proportions of the phases in the four smallest size fractions (< 2.0 µm diameter) by solving a series of simultaneous equations. This allowed a comparison to be made with the mineralogy obtained from X-ray diffraction analysis.

The methods used were sufficiently sensitive to detect small compositional variations within mineral groups. Detailed analysis of muscovite and illite grains established definite trends in the proportions of SiO_2, Al_2O_3, Fe_2O_3, and most significantly K_2O, which were a direct consequence of grain size fluctuations and were not merely an artefact of the method or a consequence of the operating conditions.

This microanalytical technique has proved to be a valuable tool in the investigation of natural dispersions of clay and clay-sized minerals, providing data which might otherwise be unobtainable using more standard methods.

ACKNOWLEDGEMENTS

The author wishes to thank Dr F. H. Hubbard and R. J. McGill for their invaluable advice and two anonymous reviewers for suggesting improvements to the manuscript.

REFERENCES

Beaman, D. R., and File, D. M., 1976, Quantitative determination of asbestos fibre concentrations, Anal. Chem., 48: 101.

Cliff, G., and Lorimer, G. W., 1975, The quantitative analysis of thin specimens, J. Microscopy, 103: 203.

Dymond, J., and Eklund, W., 1978, A microprobe study of metalliferous sediment components, Earth Planet. Sci. Lett., 40: 243.

Fitzgerald, A. G., and Storey, B. E., 1979, Modification of an EM6G electron microscope for X-ray microanalysis, J. Microscopy, 115: 73.

Giese, R. F., 1971, Hydroxyl orientation in muscovite as indicated by electrostatic calculations, Science, N.Y., 172: 263.

Kershaw, P. J., 1980, "An Investigation of Factors Influencing the Concentration of Trace Metals in the Bottom Sediments of the Forth Estuary", PhD Thesis, University of Dundee.

Mackenzie, R. C., 1975, The classification of soil silicates and oxides, in: "Soil Components. Vol. 2 Inorganic Components", J. E. Gieseking, ed., Springer-Verlag, Berlin, Heidelberg, New York.

Mackenzie, R. C., and Milne, A. A., 1953, The effect of grinding on micas. I Muscovite, Min. Mag., Lond., 30: 178.

McGill, R. J., 1980, "A Study of the Application of Electronic Microscope Microanalysis Techniques to Mineralogical Investigations", MSc Thesis, University of Dundee.

McGill, R. J., and Hubbard, F. H., 1981, Thin film k-value calibration for low atomic number elements using silicate mineral standards, in: "Quantitative Microanalysis with High Spatial Resolution", G. W. Lorimer, M. Jacob and P. Doig, eds., Metals Society, London.

Nagelschmidt, G., 1938, On the atomic arrangement and variability of the members of the montmorillonite group, Min. Mag., Lond., 25: 140.

Norrish, K., 1972, Factors in the weathering of mica to vermiculite, Proc. Int. Clay Conf., Madrid, 417.

Pearson, M. J., 1978, Quantitative clay mineralogical analyses from the bulk chemistry of sedimentary rocks, Clays Clay Miner., 26: 423.

Pooley, F. D., 1977, The use of an analytical electron microscope in the analysis of mineral dusts, Phil. Trans. R. Soc. Lond., A, 286: 625.

Read, S. J. B., 1975, "Electron Microprobe Analysis", University Press, Cambridge.

Reichenbach, H. G. von, and Rich, C. I., 1975, Fine-grained micas in soils, in: "Soil Components. Vol. 2 Inorganic Components", J. E. Gieseking, ed., Springer-Verlag, Berlin, Heidelberg, New York.

Rubin, I. B., and Maggiore, C. J., 1974, Elemental analysis of asbestos fibers by means of electron probe techniques, Environ. Hlth Perspect., 9: 81.

Slivinsky, V. W., and Ebert, P. J., 1972, $K\beta/K\alpha$ X-ray transition-probability ratios for elements $18 \leq Z \leq 39$, Physical Review A, General Physics, 5 (4): 1581.

Thompson, G. R., and Howe, J., 1975, The mineralogy of glauconite, Clays Clay Miner., 23: 289.

Wilson, M. J., 1971, Clay mineralogy of the Old Red Sandstone (Devonian) of Scotland, J. Sedimen. Petrol., 41: 995.

Wilson, M. J., Bain, D. C., McHardy, W. J., and Berrow, M. L., 1972, Clay mineral studies on some Carboniferous sediments in Scotland, Sedim. Geol., 8: 137.

ESCAPE OF PORE FLUID FROM CONSOLIDATING SEDIMENT

G. C. Sills and K. Been*

Department of Engineering Science
University of Oxford, Oxford, UK
*Now with Golder Associates, Canada

INTRODUCTION

Cohesive sediment can be moved substantial distances as suspended solids in rivers, estuaries and coastal waters, until eventually insufficient energy is available to keep it in suspension and it is deposited. It may then consolidate and form an increasingly dense stable layer, or it may be eroded and moved before settling again. During this process of erosion, movement and deposition, the sediment may encounter a wide range of physical and electrochemical conditions, causing corresponding changes in its flocculation characteristics and its behaviour. These changes do not end with deposition, as biological, bacterial and chemical action can substantially affect the consolidating sediment.

The natural sediment also provides a transport mechanism for substances artificially discharged into the water, such as industrial effluent and radioactive waste, which can become attached to the soil particles or flocs. The strength of this bonding will be determined by factors such as the floc structure and the electrochemical environment, so that these pollutants may move within the consolidating sediment, between the soil and the pore water. Thus, if there is to be any possibility of predicting the subsequent movement of these pollutants, it will be necessary to predict the strains occurring within the sediment and the movements of the pore water. The consolidation process can last for months or even years, depending on the addition or erosion of further sediment, the permeability, gas generation and the sediment itself. Initially, the sediment/water interface settles quickly and large amounts of pore water can be expelled from the soft soil. Later, the rates reduce but, over a long period of time, a substantial amount of pore fluid can be released back into the water column. If a particular pollutant is deposited with the cohesive sediment and then released into the pore fluid due to changing conditions, it would move upwards, perhaps to be redeposited on the surface sediments on encountering suitable conditions. This would lead to an increasing pollutant concentration in the surface layers. Alternatively, if the pollutant is expelled from the bed with the pore fluid, a new cycle of deposition and expulsion could start elsewhere.

Therefore, in addition to an understanding of the chemical and biological processes involved, waste disposal policies must be based on an appreciation of the sediment behaviour. This paper describes some laboratory measurements made during the deposition and consolidation of a suspension of sediment in water.

LABORATORY SIMULATION

In the field, where the energy in the water column is sufficiently high, cohesive sediment is frequently transported in suspension before being deposited in one location. Once deposition has occured, further movement takes place vertically, as the sediment packs closer together, breaking down floc structures and expelling pore water upwards in a process of consolidation. This one dimensional process is modelled in the laboratory by allowing a dilute soil slurry to settle in 102 mm diameter, 2 m high columns. At the start of an experiment, a well-mixed slurry is placed in the column at a uniform density, sufficiently low that the sediment is entirely supported by the water, with no soil framework existing. If the pressures of the water in the slurry are measured, they are found to equal the total pressure due to the weight of sediment and water. As the sediment settles, the flocs are brought closer together and begin to interact, forming a framework that helps support the total pressure. The pore water pressure drops below its initial total pressure value and water flows upward through the sediment. (It should be noted that the direction of this water flow is upward both in absolute terms - referred to stationary external co-ordinates - and in relative terms, by comparison with the sediment.) The sediment will settle until the solid particles are supported entirely by the framework that has developed in the soil. At this stage, the pore water pressure only has to balance the weight of the water, and therefore is in hydrostatic equilibrium, and the settlement is complete.

ESCAPE OF PORE FLUID FROM CONSOLIDATING SEDIMENT

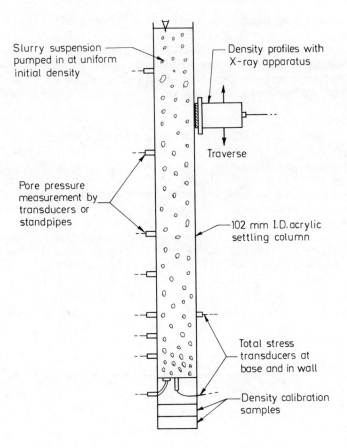

Fig. 1. Diagram of settling experiment.

The deposition from a suspension and the subsequent consolidation are monitored by measurements of density and pore water pressure throughout the column, and settlement of the sediment surface.

A high accuracy of density measurement has been achieved using an X-ray absorption technique (Been 1980, 1981). A collimated beam of X-rays, is directed towards the settling column. A proportion of the rays that pass through the sediment enters a collimated detector assembly (consisting of a sodium iodide crystal and photomultiplier) where a count rate is generated. The system is shown schematically in Fig. 1. The count rate is exponentially related to the sediment density so that, by an appropriate calibration, the density at any point in the column may be obtained. The accuracy of the system is potentially very high: a stationary measurement of density over a period of a few seconds will be accurate to ± 0.005 gm/cc with a spatial resolution of less than 1 mm. However, for these present experiments, the X-ray head traverses vertically up and down, so that a complete density profile is obtained in 1-2 minutes with an accuracy of 0.01 gm/cc and spatial resolution of 10 mm.

Measurement of pore water pressure in the settling column is achieved by pressure transducers mounted behind water-saturated porous elements set into the column wall. These porous elements are flush with the inside face of the column wall, and therefore present no obstacle to the sediment movement, but allow transmission of water pressure to the transducer, with negligible flow of water.

A pressure transducer is placed in the base of the settling column to investigate possible wall friction effects. Since no material is added or removed during an experiment, this transducer should record a constant total pressure. A reduction

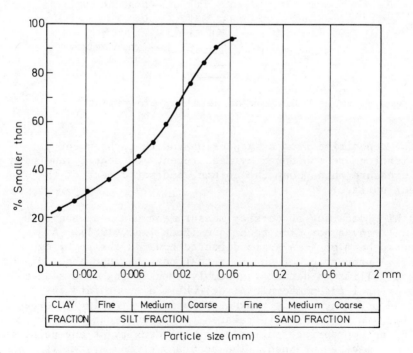

Fig. 2. Particle size analysis of Combwich mud.

from the initial value could indicate that friction has developed between the column walls and the consolidating sediment.

The sediment used for the laboratory measurements was a clayey silt from the mouth of the River Parrett, a tributary of the River Severn in the south-west of Britain. It was wet-sieved through a 75 μm sieve, using tap water. Fig. 2 shows the particle size distribution of the dispersed material.[1] However, in the natural state and in the laboratory experiments the sediment is flocculated, so the particle size distribution is a guide to classification only, not to settling behaviour. The sediment was then made up to the required density and thoroughly mixed by circulating through a pump for 10 minutes before introduction to the settling column. The flocculated sediment was pumped into the column through a pipe whose outlet was initially placed at the bottom of the column and was raised steadily to remain always just below the slurry surface. Great care was taken to avoid entrainment of air and unnecessary turbulence.

RESULTS

The transducer in the base of each settling column recorded a small drop in total pressure during the first few hours of many of the experiments. However, this appears to be associated with a small degree of particle segregation, as some of the larger silt size particles drop through the flocs to form a shallow, denser layer at the base of the column, and no further detectable changes occur during the experiment. The friction effects on the settled mud consolidation therefore seem to be negligible.

As the X-ray head moves, its position is automatically recorded against the corresponding count rate on an X-Y plotter. Fig. 3 shows typical experimental results where the x-axis, linear in the count-rate, has been marked with the corresponding unit weight value. Since the relationship between count-rate and unit weight is exponential, the unit weight scale is not linear in this representation. The results shown in Fig. 3 refer to a tap water slurry of initial unit weight 10.7 kN/m^3, (initial density 1.09 gm/cc, initial concentration 145 gm/l). The first curve is the 20-minute profile, showing that the suspension has an initial density that is uniform throughout the column, with a small increase at the bottom where a soil is forming and also some of the larger silt-size particles have dropped through. With the passing of time, the surface of the suspension moves downward and its density reduces. This drop in density is partly explained by the small amount of particle segregation that occurs (the flocs prevent much segregation), but appears to be also a genuine dilation, depending on the initial density. The soil

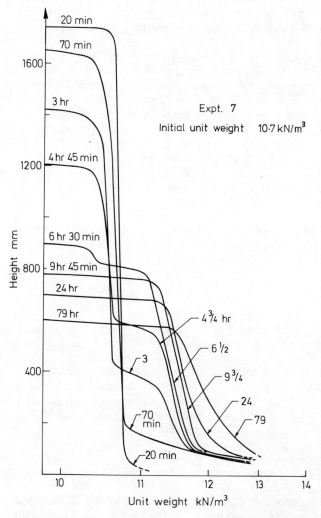

Fig. 3. Density profiles, Experiment 7. Initial unit weight 10.7 kN/m³.

layer at the bottom increases with density, while another region, intermediate between the soil and the suspension, is forming and increasing in thickness. This pattern of step changes in density is seen also in in-situ density measurements of sediment suspensions in estuaries such as the Bristol Channel, reported by Parker and Kirby (1977).

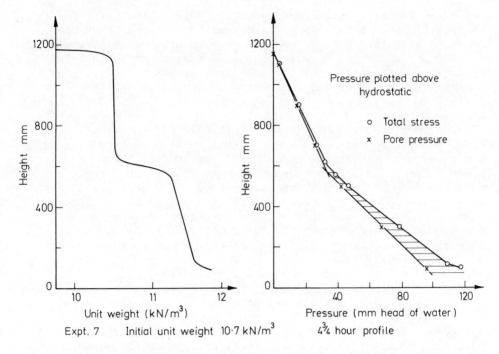

Fig. 4. Density and stress distribution. Experiment 7.

In Fig. 4, the $4\frac{3}{4}$ hour density profile is taken separately, and presented with measurements of pore water pressure. As already outlined, the pore water pressures will have a maximum value equal to the total pressure and will eventually dissipate to the hydrostatic values. Fig. 4 shows the excess pore water pressure, defined as the difference between the actual pore pressure and the equilibrium hydrostatic value at the same position. Thus the positive excess pore pressures shown in Fig. 4 indicate that consolidation is not yet complete. Fig. 4 also shows the total stress, or total pressure, distribution through the settling column, which is obtained from the density curve by calculation of the area beneath the curve. In order to make a direct comparison with the pore pressure curve, the hydrostatic pressure has been subtracted from these total stress values also. A change in physical behaviour can be seen at the transition from the initial suspension to the intermediate region: both the excess pore pressures and total pressures show a change in slope but, more importantly, in the suspension, the pore pressure and total pressure are equal, while in the intermediate region, the pore pressures are significantly lower than the corresponding total pressure. Thus, the sediment in the suspension is entirely fluid-supported, while in the intermediate region a sediment structure or framework has begun to develop. It can be seen from Fig. 3 that the density of this intermediate region increases steadily with consolidation, accompanied by an increasing density gradient through its thickness. This intermediate region would be expected to show some resistance to erosion, increasing with consolidation.

Previous measurements in the field have led to various definitions of the state of a sediment. These have been largely intuitive, and have included resistance to erosion. For example, Parker & Kirby (1977) have suggested that stationary suspensions settle to form fluid mud, with little resistance to erosion, with further development into settled mud which may be expected to remain on the sea-bed. However, the resistance to erosion is determined by the energy available as well as by the condition of the sediment, so that this criterion leads to a terminology that is not universally consistent. New definitions are therefore proposed on the basis of the physical measurements in the laboratory. Thus a suspension is seen to be defined by the absence of interparticle forces, so that the pore pressures and total stresses are equal and the sediment is entirely fluid supported. This need not be a stationary condition: the individual sediment particles or flocs may be moving in any direction. (The movement in the laboratory was downward, but water flow in the field could be lateral.) The intermediate region seen in the laboratory experiment indicates a fundamental physical change from the suspension, as a structural framework has now developed, and it is proposed to refer to this condition as settled mud. Comparing these definitions with earlier ones, the suspension, as now defined, offers no resistance to erosion, while, in any specific field situation, the settled mud could be sub-divided into erodible and non-erodible, depending solely on the energy available to cause erosion.

The changes in altitude of various density values during the first 20 hours can be seen in Fig. 5. The line with circles indicates the surface settlement, which shows a marked change in rate once the initial fluid-supported suspension has disappeared. The water/sediment interface is then at the top of the intermediate region, or settled mud, which continues to consolidate, but with a much slower settlement rate than the suspension. The other lines in Fig. 5 indicate the evolution of lines of equal density, with a marked movement upward from the base of the column while material is still being added to the settled mud layer as the suspension is deposited. At the end of this stage, the equal density lines move upward more slowly as the settled mud generally becomes more dense with consolidation.

ESCAPE OF PORE FLUID FROM CONSOLIDATING SEDIMENT

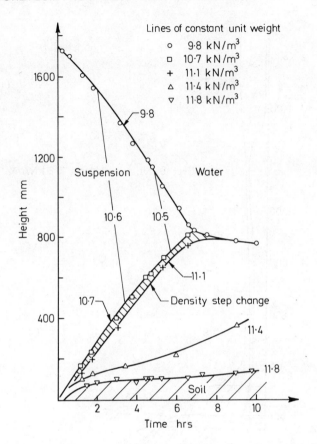

Fig. 5. Surface and internal settlements. Experiment 7.

The same qualitative behaviour is seen for a range of initial unit weights up to about 11.2 kN/m^3, (initial densities 1.14 gm/cc, initial concentrations 225 gm/1) and is shown in Fig. 6 for experiment no. 10, with initial unit weight 10.0 kN/m^3, (initial density 1.02 gm/cc, initial concentration 30 gm/1). It can be seen that the lower initial density has caused the step to the intermediate region or settled mud to occur at a slightly lower density also. Since the final, settled density distribution should depend only on the mass of sediment in the suspension and not on its initial density, this implies that larger strains will occur during the settled mud consolidation when the initial suspension density is low, since the settled mud then exists, by definition, over a wider density range.

In the present context, the knowledge of the pore water movement is of interest, and this information can be obtained

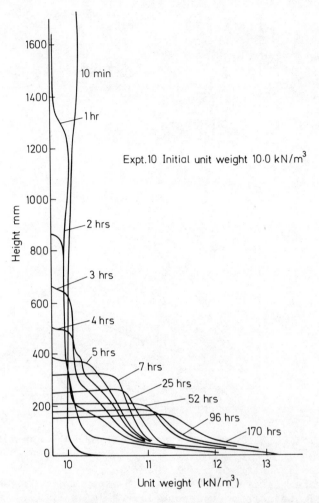

Fig. 6. Density profiles, Experiment 10. Initial unit weight 10.0 kN/m^3.

from the density profiles or surface settlement. The relevant region is the settled mud where, under field conditions, the structure of the sediment provides resistance to erosion, so that the sediment movement is mainly vertical, due to consolidation. The early stages of the laboratory settling process, during which the settled mud is established by deposition of sediment from the suspension are therefore neglected, and in the subsequent results, time t = 0 coincides with the disappearance of the suspension.

ESCAPE OF PORE FLUID FROM CONSOLIDATING SEDIMENT

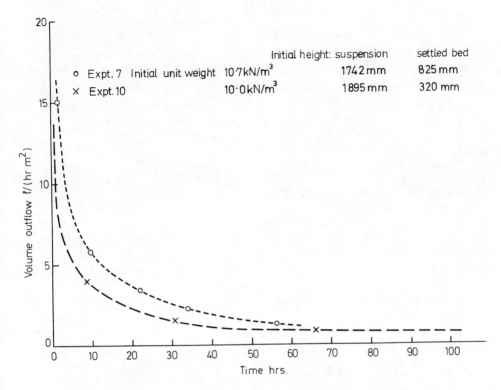

Fig. 7. Flow from surface of settled mud or intermediate region. Experiments 7 and 10.

The fluid flow across the sediment/water interface may be calculated from the rate of surface settlement. That is, if the surface has settled a distance Δh in a time interval Δt, then that height Δh is occupied by water expelled from the sediment and the rate of outward flow is Δh/Δt. A flow curve is shown in Fig. 7 for experiments 7 and 10. It can be seen that the flow rate is initially quite high, decreasing as consolidation proceeds. The actual numbers produced in these experiments are not necessarily comparable with each other, since the initial height of the settled bed is markedly different (due to the initial suspension densities being different while the initial suspension heights are similar). A better understanding of the physical behaviour may be obtained by a consideration of proportional changes during consolidation. In particular, it may be more useful to know how the pore water volume changes as a proportion of the volume initially contained in the unconsolidated settled mud. From the known initial uniform density, the ratio of water or void volume to solid volume can be

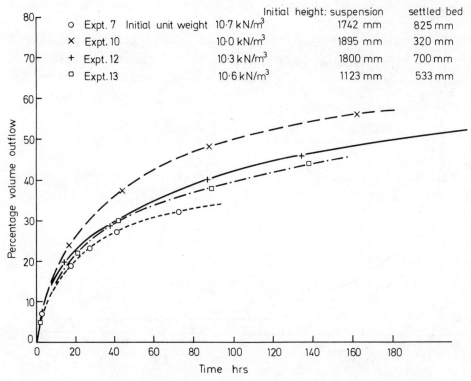

Fig. 8. Flow from surface of settled mud as a proportion of water volume initially contained in the intermediate region.

calculated from the relationship,

$$\text{void ratio } e = \frac{\rho_s - \rho_m}{\rho_m - \rho_w}$$

where ρ_s and ρ_w are the unit weights respectively of the soil particles and of water, and ρ_m is the unit weight of the initial mud slurry. Subsequent changes in void ratio are monitored from the surface movement, and the void ratio corresponding to the end of the suspension phase is then taken as the reference value. Thus, this is a measure of the volume of water initially enclosed by settled mud. Fig. 8 shows the subsequent flow of water from the settled mud as a proportion of this initial volume for four

ESCAPE OF PORE FLUID FROM CONSOLIDATING SEDIMENT

experiments, nos. 7, 10, 12 and 13. The first three all started at fairly similar initial heights in the settling column (between 1742 mm and 1895 mm) and span a range of initial unit weights from 10.0 kN/m^3 to 10.7 kN/m^3, (initial densities 1.02 gm/cc to 1.09 gm/cc, initial concentrations 30 gm/l to 145 gm/l).

As suggested earlier, it can be seen that larger strains and hence higher proportions of pore water expulsion are occurring for the lower initial densities. Of the experiments shown, only experiment 12 was continued for longer than 170 hours, although this cannot be shown in Fig. 8 due to the choice of scale. Experiment no. 12, with an initial suspension unit weight of 10.3 kN/m^3 (initial density 1.05 gm/cc, initial concentration 80 gm/l) lasted 1843 hours, by which time 67% of the original pore water had been expelled, compared with the 47% shown after 134

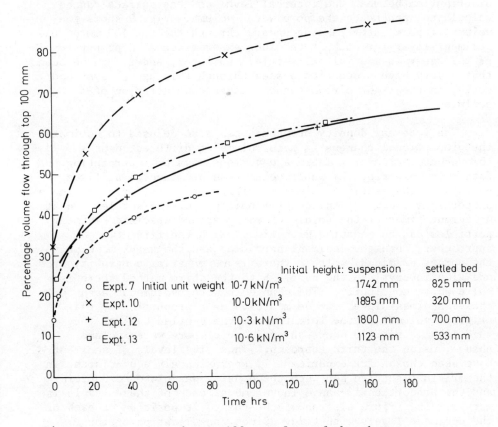

Fig. 9. Flow through top 100 mm of settled mud as a proportion of water volume initially contained in the intermediate region.

hours. Similar increases could be expected in the other cases, so that it is clear that substantial proportions of the pore water originally trapped in the settled mud will subsequently escape. The two experiments with similar initial unit weight but different initial heights, and therefore different masses of sediment, show that there is some effect due to mass. However, the data are insufficient to draw any firm conclusions.

Since the water flow considered so far has been across the water/sediment interface, a knowledge of the internal density structure has not been essential, although the initial suspension density has been required. However, in circumstances where a pollutant may be redeposited on the sediment as the pore fluid nears the surface of the settled mud, the flow is required across the appropriate internal boundary, and this can be obtained only by examination of the internal structure. Thus, the void ratio distribution between the internal level and the surface can be calculated, and hence the pore water volume. Fig. 9 shows the volume of pore water passing upward through the top 100 mm of the sediment layer for each of the four experiments as a proportion of the water in the initial settled mud layer, and it can be seen that a very high proportion passes through this upper layer and could, in the right circumstances, cause concentration of a pollutant.

The internal density structure can also be used to estimate the proportional changes in pore volume at different depths throughout the layer. There are various ways of interpreting the data - for example, the water volume can be measured at fixed heights on the settling column at different times, or at the same proportions - e.g. at the quarter points - of the settled mud at different times or the volume of water around specific sediment particles can be measured with time. With the first two approaches, both the sediment particles and the water move past the points examined, so that there is no material continuity from one time interval to the next. With the third method there is solid phase continuity. The change in water volume at each chosen sediment level can be measured as a proportion of the original volume (at the formation of the settled bed) at that same level. Fig. 10 shows an estimate of pore water volume change, using the third approach. Thus five levels of sediment have been chosen for experiment 7, approximately equi-distant in the original settled bed when the density was close to uniform and the proportional change in volume at each of these levels is marked. The figure also shows the change in position of each of the sediment layers with time. Since consolidation of the sediment occurs from the bottom upwards (i.e. the denser layers are at the bottom) the lower levels show larger proportional changes early while the upper levels initially settle without

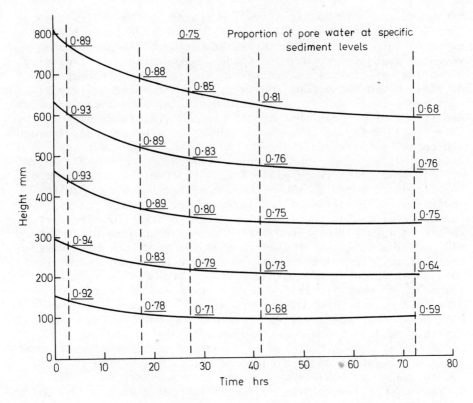

Fig. 10. Reduction in pore water volume as a proportion of the volume in the initial settled bed.

much change in density. The figures shown are probably accurate only to ± 0.03, but nevertheless indicate the general trend reasonably well.

DISCUSSION

A number of general conclusions may be drawn from the results presented here, although care must be taken in applying the results quantitatively to field conditions, where the constraints on behaviour will be more complex. However, the simplicity of the laboratory conditions makes it easier to identify some fundamental aspects of behaviour. In the absence of sediment deposition, the sediment/water interface moves downward, with sediment particles taking the place of water expelled from the pores. Pore water movement is vertically upward, regardless of whether the point of reference is taken to be the base of the column (fixed in space) or the sediment/water interface (moving in space). If the settled bed is initially unconsolidated i.e. a substantial amount of sediment has been recently deposited in a short space of time - then the velocity of pore water flow is initially higher from the bottom of the settled mud than the top. However, this situation quickly changes as the weight of the sediment causes consolidation and a density gradient develops through the bed. At later stages, the water flow is faster through the upper parts of the bed, where the porosity is greater.

If deposition takes place steadily, then the upper part of the deposited bed will always have a higher porosity than the lower and, at any given instant, the velocities will be higher in the upper part than the lower. However, since the lower part has been longer in place, the overall reduction in pore water volume will be greater at the bottom than at the top. The pore water movement will be upward everywhere through the bed when referred to a point fixed in space (the bottom of the bed or settling column). However, it is possible for the sediment/water interface to move upwards faster than the pore water at the bottom, so that the relative distance between them increases. (In this sense, the pore water movement relative to the sediment/water interface can be described as downward, although it is important to realise that the actual water flow is still upwards, against gravity). Whether this happens and, if so, where the relative velocity changes direction, will depend on the sedimentation rate, the thickness of the existing layer and the sediment properties and stress-strain behaviour. The easiest way to model this situation analytically is to work in terms of material, or Lagrange, co-ordinates, referred to a fixed point in space (see e.g. Lee & Sills 1981) rather than space, or Euler, co-ordinates.

These laboratory experiments do not provide quantitative results that can be applied directly to field situations, although it should be possible to obtain broad estimates of strain magnitude and pore water movements in cases where, due to tidal or seasonal variations, there exist periods of high deposition rates alternating with little or no sedimentation. The actual values of strain magnitude and pore water movement will be affected by the mass of sediment present and the environmental conditions. If an accurate estimate were required for a specific event, then a suitable laboratory test could be designed, making sure that the salinity and composition of the water and the sediment matched the field conditions. Measurements of density structure would allow pore water movement to be monitored.

There is also the possibility of applying some simple theories to predict the density development. Been (1980) and Been and Sills (1981) have shown that the density profiles obtained in laboratory experiments can be predicted after soil parameters have been obtained from a few experiments, using a modification to a theory developed by Lee (1979) and Lee and Sills (1981). It should also be possible to use this theory for a variable sedimentation rate, although this has not yet been tested.

CONCLUSIONS

The availability of accurate density measurement under laboratory conditions has allowed the examination of sediment settlement and consolidation. It is apparent that substantial volumes of water flow out of the sediment at rates that are dependent partly on density and that this flow and the associated settlement can occur over a time scale of months. If quantitative data are required for prediction, then a reasonably accurate simulation of the sediment and the environmental conditions would be necessary. Certainly, in any study involving water and sediment as transport mechanisms for other elements, the process of consolidation is of fundamental importance.

REFERENCES

Been, K., 1980, Stress strain behaviour of a cohesive soil deposited under water. D. Phil. Thesis, University of Oxford.
Been, K., 1981, Soil density measurement in the laboratory using X-rays. ASTM Geotechnical Testing Journal.
Been, K., and Sills, G.C., 1981, Self weight consolidation of soft soils. Geotechnique 31, No. 4, 519.
Lee, K., 1979, An analytical and experimental study of large strain soil consolidation. D. Phil. Thesis, University of Oxford.
Lee, K., and Sills, G.C., 1981, The consolidation of a soil stratum, including self-weight effects and large strains, Int. J. Num. & Analyt. Methods in Geomechanics, 5:405.
Parker, W.R., and Kirby, R., 1977, Fine sediment studies relevant to dredging practice and control. 2nd Int. Symp. on Dredging Technology, Texas.

THE INFLUENCE OF PORE-WATER CHEMISTRY ON THE BEHAVIOUR OF TRANSURANIC ELEMENTS IN MARINE SEDIMENTS

B. R. Harvey

Ministry of Agriculture, Fisheries and Food
Directorate of Fisheries Research
Fisheries Laboratory
Lowestoft, Suffolk, UK

INTRODUCTION

Marine sediments form a natural sink for many substances which would otherwise act as pollutants in the marine environment. The fate of such substances after burial depends, however, not only on the chemical nature of the substances themselves but also on the characteristics of the depositional environment with which they become associated. It has long been realized that the fluids occupying the interstices between the sediment particles play an important role in bringing about the post-depositional changes that occur within the sedimentary environment (Siever et al., 1965; Brooks et al., 1968; Friedman et al., 1968; Bischoff and Ku, 1970). Furthermore, pore-fluid is now seen as a significant reservoir and participant in the marine geochemical cycle (see Manglesdorf et al., 1969) and it must be concluded that the ultimate fate of those polluting substances which become associated with particulate matter will thus be strongly influenced by the chemistry of sediment pore waters.

Early attempts to investigate the fate of the transuranic elements, which are now finding their way into the seas via atmospheric fall-out from nuclear weapons testing or as a consequence of the discharge of low-level aqueous wastes from nuclear facilities, indicated strong affinity for sedimentary material in the case of plutonium (Pu) and americium (Am) (Hetherington, 1976). As a result of this observed affinity, considerable effort has been expended in trying to investigate the behaviour of these and other related elements within sediments after burial with a view to establishing the

© Crown Copyright

extent to which post-depositional changes might subsequently return these materials to the water phase as part of the geochemical cycling process (Hetherington, 1978; Bowen et al., 1980; Noshkin and Wong, 1980a, b; Santschi et al., 1980; Nelson and Lovett, 1981). It is perhaps unfortunate that the eastern Irish Sea, which is an area where the concentration of transuranic elements should make their behaviour in sediment easier to study than in any other part of the world, is also a coastal site. Manheim (1970) has pointed out that in such inland sea areas complex environmental factors tend to cause pore-water composition to exhibit large variations in space and time. Despite such potential difficulties it was considered essential to try to make measurements of both transuranic element concentrations and the general physico-chemical conditions occurring in the interstitial waters of the muddy sediments close to the point of discharge of aqueous waste from the nuclear reprocessing plant at Windscale in Cumbria in order to establish whether or not post-depositional changes would be likely to render these elements more or less firmly attached to the solid phase.

Encouragement to pursue such studies arose as a result of the observation by Nelson and Lovett (1978) that Pu exists in both a higher and a lower oxidation state in the overlying waters of the Irish Sea. As is shown in Table 1 these different oxidation states of Pu show markedly different affinities for the aqueous and solid phases and it is not difficult to see that a change from one oxidation state to the other would have important implications for the potential mobility of the element within a given environment (Edgington, 1981). The distribution ratio (Kd) for some other related elements between sea water and sediment are also given in the table, for reference. These have either been measured, as in the case of Pu, Am and Np, or are calculated approximately from published data as in the case of Th, U and Nd. It can be seen that large differences exist between the Kd values for the higher and lower oxidation states of Pu whilst those for the same oxidation state of different elements are quite similar.

Redox conditions in marine sediments are invariably somewhat lower than those in the overlying water column and even though the vast majority of bottom sediments can be classed as 'oxidizing' the redox conditions within the sediment regime may be such as to stabilize a different oxidation state from that which is stable in the overlying water. During the development of this work consideration has been given to the behaviour of the naturally-occurring elements U and Th whose present-day distribution can, unlike the artificially produced transuranic elements, be seen to represent their behaviour over geological time-scales including the development of various isotopic dis-equilibria (Somayajula and Church, 1973). Though Th seems only to exist in one oxidation state, U displays both a higher and lower oxidation state under suitable environmental conditions.

Table 1. Distribution Ratio (Kd) for Various Elements Between Suspended Particulate Matter and Sea Water

Element	Oxidation State	$Kd \left(\dfrac{\mu g \cdot kg^{-1} \text{ particulate}}{\mu g \cdot kg^{-1} \text{ sea water}} \right)$	Source Reference
^{232}Th	IV	$> 10^7$	a
U	VI	$\sim 10^3$	Borole et al., 1982; Harvey (unpublished)
^{237}Np	V	$\sim 5 \times 10^3$	Pentreath and Harvey, 1981
$^{239/240}$Pu	III or IV V or VI	2.5×10^6 5×10^3	Nelson and Lovett, 1978
^{241}Am	III	2.3×10^6	Pentreath et al., 1980
Nd	III	2×10^6 to 1×10^7	a

[a] Calculated from data given in Burton (1975), Mason (1966) and other literature.

The purpose of this paper is to illustrate the potential value of studying pore-water chemistry in order to investigate the behaviour of transuranic elements in marine sediments, especially the likelihood of remobilization. With the extremely limited data at present available, an attempt is made to relate the change in Pu speciation across the interface between the overlying water and the sediment with the change in redox conditions which have been measured in a small number of Irish Sea cores. The likely behaviour of Np in sediments is discussed in the light of the results of laboratory experiments since as yet no suitable environmental data exist, and the opportunities for remobilization of tri-valent actinides in anoxic sediments is considered with reference to the known behaviour of the lanthanide elements and evidence from measurements of ^{241}Am in anoxic sediments.

PHYSICO-CHEMICAL CONDITIONS IN PORE WATERS

Temperate open sea areas are in general well oxygenated throughout the entire depth of water. Below this, oxidized sediments normally occur and these oxidized sediments often persist well down into the bed before anoxic conditions develop. Organic matter is the chief source of oxygen demand and the extent to which reducing conditions develop is governed largely by the organic content of the sediment and the rate at which oxygen can be introduced from

the overlying water column. Bacteria appear to be the principal
dynamic agents which affect the redox condition of sedimentary
environments (Zobell, 1946; Sorensen, 1978) and in some areas even
the surface sediments and overlying water can become anoxic at certain seasons of the year if not at all times (Sholkovitz, 1973).
Table 2 summarizes the more important features of various redox
regimes which may exist. Typical ranges of Eh and pH are given for
the different conditions along with appropriate sources of oxygen
in each case.

Pore-water Characteristics in Oxidized Sediments

Most of the pore-water studies undertaken during the present
investigations have been made on the oxic sediments of the Irish
Sea. Reference to published literature shows however that whatever
the conditions, the confidence that may be placed in pore-water data
is particularly dependent on the care and precautions used in
sampling and squeezing (Shishkina, 1964; Bischoff and Ku, 1970).
In situ samplers as developed for example by Sayles et al. (1976)
and Ridout (1981), whilst appearing to offer solutions to many of
the problems, do not have the capability in their present forms of
producing the multi-litre samples of pore-water necessary to study
the chemistry of transuranic elements in these fluids. For the
present work, cores were obtained on the MAFF research vessel
"Cirolana" using either a 1 m x 10 cm diameter gravity corer or a
40 cm Reineck box corer from which 10 cm diameter subsamples were
subsequently taken for analysis. Immediate shipboard extrusion and
sectioning was carried out, exposed surfaces of the core were covered
with a sheet of polythene during cutting and the outside portion of
each section (about 5 mm all round) was discarded to guard against
contamination by surface water. Sections for analysis of physico-
chemical conditions were then transferred immediately to a nitrogen-
filled glove box in which they were squeezed by hydraulic pressure
to expel the pore fluids. The squeezing pots were equipped with a
0.45 µm millipore filter and pre-filter on the outlet to ensure the
removal of all sediment particles. The procedures used were basi-
cally similar to those described by Somayajula and Church (1973) and
Presley et al. (1967). An in-line platinum redox probe provided a
continuous record of the Eh of the expelled fluid; the Eh dropped
rapidly during the early period of squeezing as oxygen trapped in
the filter and other parts of the apparatus was expelled (Troup et
al., 1974). The sample of pore water on which further analyses were
to be carried out was collected once the Eh had dropped to a constant
value. Fig. 1 summarizes the measurements that have been made on
the pore water from a core taken not far from the Windscale effluent
discharge point. It shows the general features typical of the
'oxidized' sediments in the area; just below the surface the Eh
drops rapidly to well under +200 mV (on the hydrogen scale) whilst

Table 2. Typical Redox Regimes in a Marine Environment

Regime	Redox Potential mV (Hydrogen Scale)		Oxidants
Sediment/Water Interface (Oxygen not Limiting)	Eh pH	+350 to +450 mV 8.0 to 8.2	Dissolved Oxygen Others
'Oxidized' Sediments (Free Oxygen Limited)	Eh pH	+100 to +250 mV 7.0 to 7.6	$Mn^{4+} \rightarrow Mn^{2+}$ $NO_3^- \rightarrow NO_2^- \rightarrow NO^-$ etc. $Fe^{3+} \rightarrow Fe^{2+}$
Reduced Sediments (Free Oxygen Absent)	Eh pH	below zero 7.5 to 8.0	$SO_4^{2-} \rightarrow S^{2-}$ Others

the pH approaches 7.5. Bearing in mind that Eh becomes more positive by approximately 60 mV per unit decrease in pH, these readings represent appreciably more reducing conditions than those of the overlying water. As stated earlier, decomposition of organic matter within the sediments is the principal cause of the lower redox potential observed. Following the depletion of dissolved oxygen, other electron acceptors such as Mn^{4+}, NO_3^-, NO_2^-, Fe^{3+} and finally SO_4^{2-} and CO_3^{2-} are progressively used as oxidizing agents (van Kessel, 1978). It can be seen that an appreciable concentration of iron in the ferrous state is present in these interstitial waters, whilst in the sea water above the bed dissolved iron was not detected.

According to various authors (Stumm and Morgan, 1972; Theis and Singer, 1973) the reaction between iron and organic matter can be represented as:

Fe^{2+} organic complex + $1/4.O_2 \rightarrow Fe^{3+}$ organic complex;
Fe^{3+} organic complex $\rightarrow Fe^{2+}$ + oxidized organic product;
Fe^{2+} + organic matter $\rightleftarrows Fe^{2+}$ organic complex.

Depending on the relative rates of these reactions, a high steady state concentration of ferrous iron may be maintained even under comparatively oxidizing conditions provided that organic matter is present. As a result of these reactions the redox potential tends to be held comparatively steady at about +100 to +150 mV in the pH range 7.0 to 7.5 and once electron acceptors such as Mn^{4+} and NO_3^-/NO_2^- have been depleted, the ferric/ferrous system can be looked upon as virtually controlling the redox potential until the comparatively large reservoir of ferric iron in the sediments has been used up.

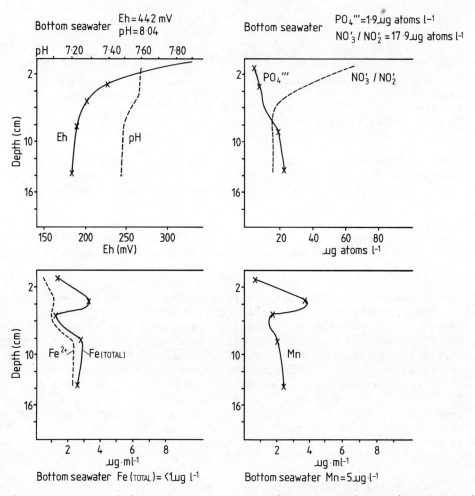

Fig. 1. Interstitial waters from an Irish Sea sediment (March 1979).

The various compounds of nitrogen present in marine sediments are interconverted by bacterial action (as part of the nitrogen cycle) but the ability of the different groups of bacteria involved to carry out their particular transformations is effectively controlled by the prevailing redox conditions (Sorensen, 1978). For the purposes of the present discussion it is important to note that significant concentrations of nitrate (up to 300 µg at.N.l^{-1}) can develop even in the presence of the substantial amounts of ferrous iron already shown to be present in this type of sediment. Furthermore, the presence of this nitrate can in effect hold the Eh of the system in the region of +200 mV (Bailey and Beauchamp, 1973) whilst it remains. The occurrence of significant, if small, amounts of nitrite (0.1 to 1% of the nitrate value in most cases) in the presence of ferrous iron can be understood only in terms of the dynamics of the complex system as a whole. The presence of this nitrite is, however, of interest with regard to the most likely oxidation state of Pu to be stabilized in these sediments. In the laboratory, nitrite would normally stabilize Pu as Pu IV whilst ferrous iron would normally cause reduction to Pu III (Coleman, 1965). The question at first seems rather an academic one but may have importance in predicting (or explaining) the extent to which Pu may form soluble complex species with organic ligands in sediments.

Behaviour of Pu and Np in Oxidized Sediments

As has been noted by Bischoff and Ku (1970) for example there is strong evidence that some elements at least are recycled through pore fluids from sediments back to sea water. It is not surprising therefore that considerable effort has been made to investigate the possibility that transuranic elements might be recycled in this way. Much, though not all, of the evidence put forward so far either to support or refute the idea of post-depositional mobility of these elements has come from analyses of sediment solids. Perhaps understandably therefore there are conflicting conclusions at the present time (Koide et al., 1975; Hetherington, 1976; Livingston and Bowen, 1979; Noshkin and Wong, 1980a, b; Aston and Stanners, 1981; Nelson and Lovett, 1981). Some of the pore-water data given in the last of these references were made in conjunction with the measurements of other parameters such as Eh, pH and ferrous iron on the "Cirolana" cruise 5/78. Whilst some of these data have already been published separately (Harvey, 1981) they have not until now been compared with the Pu data presented by Nelson and Lovett.

Portions of pore water for transuranic analysis were subjected to slightly different extraction procedures from those destined for the analysis of other parameters. Due to the large volumes required for transuranic analysis these portions were not kept under nitrogen whilst squeezing took place. There are two possible risks here: (1) Pu might be re-oxidized; and (2) the ferrous iron in solution might precipitate as $Fe(OH)_3$. The first of these is an unlikely possibility over the time-scale of the extraction process under the existing conditions. The unique characteristic of Pu is that several oxidation states can co-exist and the oxidation of Pu IV, which involves the formation of plutonium oxygen bonds (to give PuO_2^+ or PuO_2^{2+}), is known to be slow (Cleveland, 1979). The oxidation of ferrous iron on the other hand is much more likely but according to Lovett (personal communication) no precipitate of $Fe(OH)_3$ was found on the membranes used to filter the individual portions of the extracted water (150-200 ml) which went to make up the bulk sample. Some did however form in the bulked filtrates after several hours but this was redissolved before further analysis took place.

Table 3 shows the measurements of Eh, pH and the concentration of ferrous iron found in the pore-waters and the overlying sea water for three core stations in the Windscale area. Details of total $^{239/240}$Pu and the percentage of Pu in the lower oxidation state calculated from the data given in Nelson and Lovett (1981) are also given. At the sea surface the proportion of Pu in the lower oxidation state varies from 12 to 17% and only increases slightly (20-24%) just above the surface of the sediment. Immediately below the sediment/water interface almost all the Pu present is now in the reduced form. Furthermore, apart from one section (core 116, section 5-10 cm) the concentration of total Pu in the pore-water is nearly an order of magnitude lower than in the water immediately overlying the sediment. Two tentative conclusions may be drawn from these data: (1) the higher oxidation states of Pu are unstable under the redox conditions shown to exist in these cores; (2) the likely flux of Pu will be into rather than out of the bed. Such observations are in keeping with the known stability fields of the various oxidation state species of Pu thought to exist in the environment (Bondietti and Sweeton, 1977; Rai et al., 1980). It is clear therefore from these observations that the post-depositional behaviour of Pu in these and other types of sediment will be strongly influenced by the prevailing chemical conditions developed in the pore-waters.

The composition of the sediments at each of the three core stations is described as sandy mud (Pantin, 1978). Our own measurements indicate 15-20% material > 63 μm and up to 5% < 2 μm. Little variation in composition with depth was observed in any of the cores down to 30-40 cm. There is however considerable evidence of bioturbation in these sediments according to Williams et al. (1981). They consider this bioturbation to be the dominant mechanism in determining the sedimentological structure of the top 45 cm. The influence of this type of disturbance on the fluxes of Pu and other materials within the sediment and across the sediment/water interface is as yet unknown but will have to be taken into account in future studies of post-depositional behaviour in Irish Sea sediments.

Neptunium is not reduced to the trivalent state under environmental conditions. In oxidized sea water it is present predominantly as the highly soluble pentavalent oxy-cation NpO_2^+ but reference to standard electrode potentials and published redox equilibria in aqueous media (Rai et al., 1980; Inoue and Tochiyama, 1977) leads to the conclusion that Np requires a slightly lower redox potential than Pu before the tetravalent state becomes the dominant form. Attempts to measure the oxidation state of Np in Irish Sea sediment pore-water have yet to be made as this element is present in even lower concentrations than the Pu in the Windscale discharges. Laboratory experiments have, however, been carried out in simulated conditions based on the physico-chemical conditions now known to

Table 3. Percentage Plutonium in Sea Water and Pore Waters Compared with Eh, pH and Ferrous Iron Concentration

Station Number	Sample Type	Depth (cm)	Eh (mV H_2 Scale) ± 30 mV	pH	Fe^{2+} ($\mu g \cdot ml^{-1}$) ± 0.2 $\mu g \cdot ml^{-1}$	Total $^{239/240}Pu$[a] ($fCi \cdot l^{-1}$)	% Reduced Pu[b]
05/78/105	Sea Water	Surface	445	8.03	ND	185 ± 9	12 ± 1
	"	Bottom	353	7.94	"	76 ± 4	20 ± 2
	Pore Water	0– 5	335	7.26	0.3	9 ± 2	86 ± 23
		5–10	173	7.27	8.6	16 ± 2	91 ± 13
		10–15	194	7.33	4.0	15 ± 2	81 ± 12
		15–20	211	7.33	3.5	5 ± 2	62 ± 30
		20–25	176	7.46	2.6	6 ± 2	60 ± 27
		25–30	209	7.41	2.4	7 ± 2	48 ± 19
05/78/113	Sea Water	Surface	382	8.03	ND	208 ± 10	13 ± 1
	"	Bottom	384	7.84	"	79 ± 4	24 ± 2
	Pore Water	0– 5	319	7.65	0.6	14 ± 2	83 ± 14
		5–10	174	7.45	1.9	26 ± 2	77 ± 7
		10–15	181	7.43	2.7	13 ± 2	92 ± 16
		15–20	187	7.60	0.8	14 ± 2	86 ± 14
		20–25	170	7.40	2.2	16 ± 2	78 ± 12
		25–30	154	7.28	6.8	7 ± 2	72 ± 24
05/78/116	Sea Water	Surface	387	8.02	ND	377 ± 19	17 ± 1
	"	Bottom	372	7.91	"	1305 ± 70	23 ± 2
	Pore Water	0– 5	240	7.55	0.7	182 ± 9	96 ± 7
		5–10	142	7.27	3.4	572 ± 29	98 ± 7
		10–15	160	7.44	2.8	66 ± 4	95 ± 7
		15–20	194	7.41	1.0	34 ± 3	94 ± 9
		20–25	190	7.42	1.2	27 ± 2	89 ± 9
		25–30	189	7.54	0.9	16 ± 2	88 ± 13

ND – Not detected.
[a] Total $^{239/240}Pu$ from Nelson and Lovett, 1981.
[b] % calculated from data given in Nelson and Lovett, 1981.

exist in the pore water of the sediments of interest (Harvey, 1981). Using the short-lived gamma-emitting isotope ^{239}Np as a tracer, fine Irish Sea sediment material was stirred with sea water by bubbling nitrogen through the mixture in a closed reaction vessel. The amount of Np reduced from the pentavalent to tetravalent state was determined over a range of conditions covering those known to occur in oxidized sediments. Ferrous iron was used as the reducing agent and dilute hydrochloric acid was added to reduce the pH to the required point. The results of these experiments are summarized in Table 4, and comparing these results with the environmental data given for Pu in Table 3, it would appear that Np V is somewhat more resistant to reduction than is the corresponding higher oxidation state of Pu (probably Pu V). In the pH range 7.4 to 7.6 no reduction could be detected during the 8-hour experiment when the Eh was maintained between 170 and 190 mV. In the natural environment Pu was substantially reduced in the upper layers of sediments, apparently in that same range of conditions.

Clearly, Np should be investigated under environmental conditions, the importance of establishing its behaviour in marine sediments being two-fold. Firstly, ^{237}Np is one of the longest-lived, artificially-produced α-emitting radionuclides in the Windscale discharge and is of interest in connection with the very long-term environmental impact of transuranic elements, for example from the dumping of high-level solid radioactive wastes. Secondly, and related to this long-term effect, is the fact that ^{237}Np is the daughter product in the decay chain $^{241}\text{Pu} \xrightarrow[14.7 \text{ y}]{\beta'} {}^{241}\text{Am} \xrightarrow[432 \text{ y}]{\alpha} {}^{237}\text{Np}$. If, therefore, both Pu and Am are strongly bound by sediments it follows that much of the ^{237}Np produced in the environment in years to come will occur at depth within sediments. Its ability to migrate out from the sediments back into the water column, and thus to become available to the biosphere, will therefore depend to a large extent upon the oxidation state in which it is produced.

Comparison between the relative behaviour of the naturally-occurring radio elements U and Th provide confirmation of the importance of oxidation state changes on environmental mobility. Th, whose oxidation state is invariably Th IV, binds so strongly to sediments that the detection of soluble ^{232}Th in sea water is extremely difficult. U, which is present even in oxic sediments as the hexavalent UO_2^{2+} ion, shows a steady state concentration in excess of 3 µg.l^{-1} in sea water. In anoxic sediments where it does become reduced to U IV uranium becomes firmly fixed to the solid phase and naturally-occurring uranium ores were probably all laid down under such conditions. Various workers (Ku, 1965; Kolodny and Kaplan, 1970) have demonstrated the ascending migration of the radiogenic nuclide ^{234}U in oxic sediments which is produced by the decay of its parent ^{234}Th but no evidence has ever been seen of any mobility of Th within the pore-water system of marine sediments.

Table 4. Reduction of ^{239}Np in Simulated Pore-water Condition at Various Values of Eh and pH

pH Range	Eh (mV Hydrogen Scale)	% Np Reduced (Np V → Np IV)
6.9 to 7.2	180 to 200	0
	150 to 170	34
	120 to 140	72
	90 to 110	100
7.4 to 7.6	170 to 190	0
	140 to 160	67
	110 to 130	73
7.8 to 8.0	170 to 190	0
	150 to 170	11
	120 to 140	25

Pore-water Characteristics of Anoxic Sediments

Anoxic sediments are of interest in a study of the environmental behaviour of transuranics for two reasons. Firstly, by their very nature, they can be expected to provide substantially more organic compounds capable of forming complex species with transuranic elements than are likely to occur in oxic sediments and, secondly, because there is reason to suppose that the actinides in the trivalent state are known to share many chemical characteristics with the lanthanides, there is a possibility that they may follow the lanthanides in being solubilized by phosphate released in anoxic conditions. Bonatti et al. (1971) have pointed out that phosphate is released when Fe^{2+} and Ca^{2+} become bound (as a result of bacterial action) in the form of the more insoluble sulphides and carbonates respectively. The removal of these cations from solution leaves the phosphate which then causes the solubilization of suitable trivalent cations such as the rare earths. These authors have shown that lanthanum and phosphate appear to precipitate together in the oxic layers overlying anoxic sediments in the eastern Pacific. It is indeed a fact that rare earth elements generally are hosted mainly in phosphate minerals.

Transuranics in Anoxic Sediments

Neither americium (Am) nor curium (Cm) display anything other than trivalency under environmental conditions as far as is known at the present time. As mentioned earlier, Pu may exist, if only in part, as Pu III in oxic sediments, and almost certainly so under strongly reducing conditions. The possibility exists therefore that these trivalent actinides might solubilize, along with the rare-earths and phosphate, under anoxic conditions, but so far no evidence appears to have been published to support or disprove this suggestion. This is probably because very little Am, other than fall-out amounts, exists in areas where anoxic sediments have been studied for radionuclides, and in the Irish Sea, where Am could be studied comparatively easily, very few upper sediment layers are anoxic in character.

The formation of soluble complexes of transuranic nuclides with organic ligands, especially in anoxic sediments, cannot be ruled out. Theoretically these could influence the behaviour of these elements in pore water offering another potential path for the return of buried components to the surface in what might well be a more bio-available form than the purely inorganic compounds. Our own measurements of organically bound transuranics have been restricted to studies in oxic sediments. In Irish Sea sediments small but definite amounts of Np, Pu and Am have been found associated with humic substances but it has not yet been possible to detect soluble organic complexes of transuranic elements in pore waters. Alkaline extraction of the humic fraction from sediments in the laboratory has suggested that the acid-soluble fulvic acid fraction contains considerably less of the transuranics than does the acid-insoluble humic acid fraction. Nelson and Lovett (1981) have concluded from their pore-water studies that there is no evidence to suggest that a significant amount of Pu could be lost from the Irish Sea sediments to the overlying water column through all the processes involving Pu within pore waters in the samples that they investigated. Their conclusions are therefore not inconsistent with the view that organic complexes do not contribute significantly to the mobility of Pu in these sediments.

Evidence for the role of organic complexes in mobilizing transuranic elements under anoxic conditions is largely lacking though their involvement has often been assumed to be a major one. The ability of the various oxidation states to form strong bonds with complexing ligands increases in the general order V < VI < III << IV. This would support the view that those elements remaining in the tetravalent state in strongly reducing conditions, such as Np, Th and U, should show a greater tendency to form organic complexes than those existing in the trivalent state, such as Am, Cm and Pu. Clearly, there is scope for much further study in this area but a recent investigation of Pu and Am behaviour in anoxic marine sediment by Carpenter and Beasley (1981) offers evidence against remobilization. There is overwhelming evidence for the naturally-occurring elements U and Th (and the lanthanide elements) that organic complexation processes do not produce a major resolubilization of these elements in anoxic sediments.

CONCLUSIONS

The present study in which the influence of the general physico-chemical conditions existing in the pore waters of the Irish Sea sediments on the behaviour of transuranic elements has been made, indicates a dependence, either directly or indirectly, for most of the elements, on the prevailing redox conditions.

Certain features of the transuranic elements, uranium, thorium and the lanthanides, such as their partition between the aqueous and solid phases of the system and their ability to form complex species, seem to be similar for the same oxidation state of the different elements. Common features of this type are useful for interpreting observed features in different situations and provide a limited opportunity for the use of analogues to predict the likely behaviour of the transuranic elements in areas where they do not at present exist in measurable quantities. Modelling exercises for potential dumping sites in the deep ocean are an example of such a use.

Though reasonable information is now forthcoming for oxic sediments, our knowledge of the behaviour of transuranics in the pore water of anoxic sediments seems somewhat lacking. This is chiefly due to a lack of transuranic nuclides in areas where anoxic sediments occur, the concentration of these nuclides from fall-out resulting from nuclear weapons testing in the atmosphere being insufficient for suitable measurements to be made. Furthermore, there is considerable difficulty in simulating the necessary conditions in the laboratory so that meaningful experiments can be carried out.

REFERENCES

Aston, S. R., and Stanners, D. A., 1981, Plutonium transport to and deposition and immobility in Irish Sea intertidal sediments, Nature, Lond., 289:581.

Bailey, L. D., and Beauchamp, E. G., 1973, Effects of temperature on nitrate and nitrite reduction, nitrogenous gas production and redox potential in a saturated soil, Can. J. Soil Sci., 53:213.

Bischoff, J. L., and Ku, T., 1970, Pore-fluids of recent marine sediments. 1. Oxidizing sediments of 20°N, Continental rise to mid-Atlantic ridge, J. Sedim. Petrol., 40:960.

Bonatti, E., Fisher, D. E., Joensuu, O., and Rydell, H. S., 1971, Postdepositional mobility of some transition elements, phosphorus, uranium and thorium in deep sea sediments, Geochim. Cosmochim. Acta, 35:189.

Bondietti, E., and Sweeton, F., 1977, Transuranic speciation in the environment, in: "Proceedings of the Symposium Transuranics in Terrestrial and Aquatic Environments", Gatlinberg, Tennessee, 5-7 October 1976, M. G. White and P. B. Dunaway, eds., Rep. U.S. Dept. En., NVO-178:449.

Borole, D., Krishnaswami, S., and Somayajula, B., 1982, Uranium isotopes in rivers, estuaries and adjacent coastal sediments of western India: their weathering transport and oceanic budget, Geochim. Cosmochim. Acta, 46:125.

Brooks, R., Presley, B., and Kaplan, I., 1968, Trace elements in the interstitial waters of marine sediments, Geochim. Cosmochim. Acta, 32:397.

Bowen, V., Noshkin, V., Livingston, H., and Volchok, H., 1980, Fall-out radionuclides in the Pacific Ocean: vertical and horizontal distributions, largely from Geosecs stations, Earth Planet. Sci. Lett., 49:411.

Burton, J., 1975, Radioactive nuclides in the marine environment, in: "Chemical Oceanography", J. Riley and G. Skirrow, eds., 3, 2nd Edition, Academic Press, London and New York.

Carpenter, R., and Beasley, T. M., 1981, Plutonium and americium in anoxic marine sediments: evidence against remobilization, Geochim. Cosmoshim. Acta, 45:1917.

Cleveland, J. M., 1979, "The Chemistry of Plutonium", 2nd Edition, Gordon and Breach, New York.

Coleman, G. H., 1965, "The Radiochemistry of Plutonium", U.S. Atomic Energy Commission/National Academy of Sciences/National Research Council, Washington, Publication NAS-NS 3058.

Edgington, D. N., 1981, Characterization of Transuranic Elements at Environmental Levels, in: "Techniques for Identifying Transuranic Speciation in Aquatic Environments" Proceedings of an IAEA/CEC Technical Committee, Ispra, 24-28 March 1980, IAEA, Vienna, ST1/PUB/613.

Friedman, G., Fabricand, B., Imbimbo, E., Brey, M., and Sanders, J., 1968, Chemical changes in interstitial waters from continental shelf sediments, J. Sedim. Petrol., 38:1313.

Harvey, B. R., 1981, The potential for post-depositional migration of neptunium in Irish Sea sediments, in: "Impacts of Radionuclide Releases into the Marine Environment", Proceedings of an IAEA/OECD Symposium, Vienna, 6-10 October 1980, IAEA, Vienna, Publication SM 248.

Hetherington, J. A., 1976, Behaviour of plutonium nuclides in the Irish Sea, in: "Environmental Toxicity of Aquatic Radionuclides: Models and Mechanisms", M. Miller and J. Stannard, eds., Ann Arbor Science Publishers, Ann Arbor, Michigan.

Hetherington, J. A., 1978, The uptake of plutonium nuclides by marine sediments, Mar. Sci. Commun., 4:239.

Inoue, Y., and Tochiyama, O., 1977, Determination of the oxidation states of neptunium at tracer concentrations by adsorption on silica gel and barium sulphate, J. Inorg. Nucl. Chem., 39:1443.

Kessel, J. F. van, 1978, The relation between redox potential and denitrification in a water-sediment system, Wat. Res., 12: 285.

Koide, M., Griffin, J., and Goldberg, E., 1975, Records of plutonium fallout in marine and terrestrial samples, J. Geophys. Res., 80:4153.

Kolodny, Y., and Kaplan, I. R., 1970, Uranium isotopes in sea-floor phosphorites, Geochim. Cosmochim. Acta, 34:3.

Ku, T.-L., 1965, An evaluation of the $^{234}U/^{238}U$ method as a tool for dating pelagic sediments, J. Geophys. Res., 70:3457.

Livingston, H. D., and Bowen, V. T., 1979, Pu and ^{137}Cs in coastal sediments, Earth Planet. Sci. Lett., 43:29.

Mangelsdorf, P. C., Wilson, T. R., and Daniell, E., 1969, Potassium enrichment in interstitial waters of recent marine sediments, Sci. J., 165:171.

Manheim, F. T., 1970, "Interstitial Waters. Encyclopaedia of Science: Geochemistry Volume", McGraw Hill, New York.

Mason, B., 1966, "Principles of Geochemistry", John Wiley, New York and London.

Nelson, D. N., and Lovett, M. B., 1978, Oxidation states of plutonium in the Irish Sea, Nature, Lond., 276:599.

Nelson, D. N., and Lovett, M. B., 1981, Measurement of the oxidation state and concentration of plutonium in interstitial waters of the Irish Sea, in: "Impacts of Radionuclide Releases into the Marine Environment", Proceedings of an IAEA/OECD Symposium, Vienna, 6-10 October 1980, IAEA, Vienna, Publication SM 248.

Noshkin, V. E., and Wong, K. M., 1980a, Plutonium mobilization from sedimentary sources to the marine environment, in: "Marine Radioecology", Proceedings of the 3rd NEA Seminar on Marine Radioecology, Tokyo, 1-5 October 1979, OECD, Paris.

Noshkin, V. E., and Wong, K. M., 1980b, Plutonium in the north equatorial Pacific, in: "Processes Determining the Input, Behaviour and Fate of Radionuclides and Trace Elements in Continental Shelf Environments", U.S. Dept. Energy, Washington D.C., Rep. Conf. 790382.

Pantin, H. M., 1978, Quaternary sediments from the north-east Irish Sea: Isle of Man to Cumbria, Bull. Geol. Surv. Gt Br., 64: 43 pp.

Pentreath, R. J., and Harvey, B. R., 1981, The presence of ^{237}Np in the Irish Sea, Mar. Ecol., Progr. Ser., 6:243.

Pentreath, R. J., Jefferies, D. F., Lovett, M. B., and Nelson, D. M., 1980, The behaviour of transuranics and other long-lived radionuclides in the Irish Sea and its relevance to the deep sea disposal of radioactive wastes, in: "Marine Radioecology", Proceedings of the 3rd NEA Seminar on Marine Radioecology, Tokyo, 1-5 October 1979, OECD, Paris.

Presley, B. J., Brooks, R. R., and Kappel, H. M., 1967, A simple squeezer for removal of interstitial water from ocean sediments, J. Mar. Res., 25:355.

Rai, D., Serne, R. J., and Swanson, J. L., 1980, Solution species of plutonium in the environment, J. Environ. Qual., 9:417.

Ridout, P. S., 1981, A shipboard system for extracting interstitial water from deep ocean sediments, Rep. Inst. Oceanogr. Sci., U.K., 121:11 pp.

Santschi, P. H., Li, Y.-H., Bell, J., Trier, R., and Kawtaluk, K., 1980, Plutonium in coastal marine environments, Earth Planet. Sci. Lett., 51:248.

Sayles, F. L., Mangelsdorf, P. C., Wilson, T. R., and Hume, D. N., 1976, A sampler for the in situ collection of marine sedimentary pore waters, Deep Sea Res., 23:259.

Shishkina, O. V., 1964, Chemical composition of pore solutions in oceanic sediments, Geochem. Int., 3:522.

Sholkovitz, E., 1973, Interstitial water chemistry of the Santa Barbara Basin sediments, Geochim. Cosmochim. Acta, 37:2043.

Siever, R., Beck, K. C., and Berner, R. A., 1965, Composition of interstitial waters of modern sediments, J. Geol., 73:39.

Somayajula, B. L., and Church, T. M., 1973, Radium, thorium and uranium isotopes in the interstitial water from Pacific Ocean sediments, J. Geophys. Res., 78:4529.

Sorensen, J., 1978, Occurrence of nitric and nitrous oxides in a coastal marine sediment, Appl. Environ. Microbiol., 36:809.

Stumm, W., and Morgan, J. J., 1972, "Aquatic Chemistry: an Introduction Emphasizing Chemical Equilibria in Natural Waters", Wiley Interscience, New York and London.

Theis, T. L., and Singer, P. C., 1973, The stabilization of ferrous iron by organic compounds in natural waters, in: "Trace Metals and Metal-Organic Interactions in Natural Waters", P. C. Singer, ed., Ann Arbor Science Publishers, Ann Arbor, Michigan.

Troup, B. N., Bricker, O. P., and Bray, J. T., 1964, Oxidation effect on the analysis of iron in interstitial water of recent anoxic sediments, Nature, Lond., 249:237.

Williams, S. J., Kirby, R., Smith, T. L., and Parker, W. R., 1981, Sedimentation studies relevant to low level radioactive effluent dispersal in the Irish Sea, Part II, Rep. Inst. Oceanogr. Sci., U.K., 120:50 pp.

Zobell, C. E., 1946, Studies on redox potential of marine sediments, Bull. Am. Assoc. Petrol. Geol., 30:477.

THE INCORPORATION OF RADIONUCLIDES INTO ESTUARINE SEDIMENTS

B. Heaton[1] and J.A. Hetherington[2]

[1] University of Aberdeen, Foresterhill, Aberdeen, UK

[2] Scottish Development Department, Pentland House
Edinburgh, UK

INTRODUCTION

In 1974 Hetherington and Jefferies (1974) published a review of the results obtained by the Fisheries Radiobiological Laboratory in the course of their monitoring of radioactivity in sediments in the North East Irish Sea area and the coasts around the Windscale fuel element reprocessing plant. The paper dealt with the more important fission product radionuclides such as zirconium 95/niobium 95, ruthenium 106, caesium 134 and 137 and cerium 144. In addition to describing the mechanism and scale of uptake in surface sediment, the paper dealt with the depth distribution of selected radionuclides in core samples taken from estuarine and coastal sites. The distributions of activity with depth appeared to take exponential forms and for this reason a model based on Fickian diffusion was proposed for the process by which radioactivity moved from sea water into the sedimentary deposits of the area.

Although the depth profiles observed were believed to be the result of a number of processes acting together such as molecular diffusion, biological activity and sedimentation, no experimental evidence was available to permit the relative importance of each such process to be established, so a model was developed in terms of an apparent diffusion coefficient after Duursma and Gross (1971). In effect this model assumed that the mechanism of incorporation was governed by a process akin to Fickian diffusion and the profile of each nuclide could be described in terms of one simple factor defined as the apparent diffusion coefficient.

Values of this factor were derived for the important fission products.

Comparison of the values of the apparent diffusion coefficients so derived with the molecular diffusion coefficients obtained from laboratory studies showed, however, large discrepancies. For caesium, the field studies gave an apparent value of something like 3×10^3 cm^2 y^{-1} whereas laboratory work gave a molecular diffusion coefficient for this radionuclide in similar chemical form and in sediments of similar mineralogy and grain size of some two orders of magnitude less. It was concluded therefore that mechanisms other than simple molecular diffusion must be operating in the sediments to produce the profiles observed and to explain the speed with which caesium nuclides appeared in the deep layers of consolidating sediments. At the time it was suspected that the most probable explanation lay in terms of biological activity which was regarded as much more likely than sedimentation to explain the rates observed.

Subsequent work on transuranic elements however suggested very strongly that sedimentation rates in the area of the previous studies with fission products were very much greater than had been suspected and that the sedimentation of already contaminated material was probably the principal mechanism by which radionuclides were being incorporated into sedimentary deposits in the area. From these studies, sedimentation rates were derived for the area of the earlier fission product studies and with this information it has been possible to improve the earlier model of Hetherington and Jefferies (1974). In this paper details are given of a new model which takes account of sedimentation, adsorption, molecular diffusion and radioactive decay in the uptake and distribution of radionuclides in accreting sediments. Although several models have been proposed along similar lines in recent years (Aston and Stanners, 1979; Lerman, 1971) the evidence suggests that the present one is attractive in two respects. On the one hand it is simple to apply and does permit under some circumstances a full solution to be obtained for the diffusion equation as opposed to a numerical one. On the other hand it appears to be capable of modelling with precision the distributions butions observed in the environment. The implications for biological studies are briefly discussed.

THE MODEL

The general one dimensional equation for the rate of change in the concentration C of a radioactive species in solution in the interstitial water of a sediment under the influence of sedimentation, diffusion, adsorption and radioactive decay is

$$\frac{\partial C}{\partial t} = \frac{D}{1+K} \cdot \frac{\partial^2 C}{\partial Z^2} - \frac{U}{1+K} \cdot \frac{\partial C}{\partial Z} - \frac{\lambda C}{1+K} \qquad (1)$$

where U is the sedimentation rate, D the diffusion coefficient, λ the radioactive decay constant and K the distribution coefficient defined as the ratio (Radionuclide concentration on solid sediment, C_s)/(Radionuclide concentration in interstitial water, C). K is assumed to be independent of Z and t in a given core.

The general solution of equation (1) gives the concentration $C(Z,t)$ in interstitial water as a function of depth, Z, at any time, t. To be of practical value a solution must be obtained subject to boundary conditions, in particular those relating to the concentration of the radioactivity in the sea water at the time of sediment deposition. The solution has been obtained therefore subject to the following boundary conditions:

$Z = 0$ is always taken at the sediment/sea water interface.

$C(Z,0) = 0$ for all Z.

$C(Z,t) \rightarrow 0$ as $Z \rightarrow \infty$ for all t.

$C(0,t) = F(t)$ where $F(t)$ describes the variation with respect to time of the radioactivity in the overlying sea water.

Subject to these conditions the solution is

$$C(Z,T) = \frac{Z}{(1+K)2\sqrt{\pi D}} \exp\left[\frac{UZ}{2D}\right]$$

$$\times \int_0^t F(t-P) \exp\left[-\frac{(U^2 + 4\lambda D)P}{4D} - \frac{Z^2}{4DP}\right] P^{-3/2} dP$$

where t is the total time for which the process has been underway and P is an arbitrary time increment selected according to the scale of the process.

This solution has the form of the solution for an instantaneous point input multiplied by a convolution integral. It is simpler than solutions derived previously in that it does not contain multiple error functions.

Equation (2) gives the concentration in interstitial water. The concentration on the solid phase is given by C_s (Z,t) where

$$C_s (Z, t) = K C (Z, t)$$

and the concentration in bulk sediment ie the solid plus interstitial water is given by C_b (Z, t) where

$$C_b (Z, t) = (1 + K) C (Z, t)$$

Thus when considering bulk sediment, equation (1) becomes independent of K as does, therefore, the solution in terms of the variables Z and t.

APPLICATION OF THE MODEL

To test the model, it has been used to predict depth profiles of caesium 137 and plutonium 239 and 240 in areas of the Ravenglass Estuary from which cores have already been investigated experimentally (Hetherington and Jefferies, 1974; Hetherington, 1976). Plutonium 239 and plutonium 240 were measured together by the radioanalytical technique in that work and all references to plutonium 239 included plutonium 240. The same applies in this paper. The model operates essentially on three variables - the form of the radionuclide input function, the diffusion coefficient and the sedimentation rate. In one sense the radionuclide input function presents the greatest difficulty. To obtain a solution to the basic equation in analytical form requires F(t) in the form of a continuous smooth function. Although patterns of discharge of plutonium 239 and caesium 137 from Windscale with respect to time are well documented, they are not smooth and do not lend themselves to expression in analytical form. For this reason, the discharge data have been averaged on an annual basis and the equation has been solved numerically by computer for a selected set of values of Z at a value of t corresponding to the date of collection of the cores. The discharge data used have been derived from information supplied during the Windscale Inquiry (1978).

RESULTS

Fig 1 shows the profile of plutonium 239 predicted by the model for a core taken in 1974 from an intertidal area in the vicinity of Newbiggin on the south side of the Ravenglass Estuary.

As the solution of the model does not contain a factor which directly relates the discharges from the processing plant to the activity found in a core sample the measured and predicted profiles have been normalised in order to compare them. This normalisation has been undertaken by setting the maximum concentration in each profile at 100 arbitrary units and relating the concentrations at other levels to this. The core was investigated previously by Hetherington (1976) who deduced from the variation with respect to depth of the ratio of plutonium 239 to plutonium 238 a sedimentation rate of 2.3 cm y^{-1}. Using this value for the sedimentation rate as a guide a value of the dif-

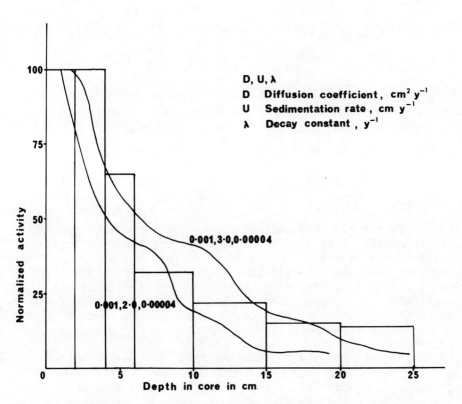

Fig. 1 Predicted distribution of Pu 239+240 for core taken from Newbiggin area of the Ravenglass Estuary and the observed profile in histogram form.

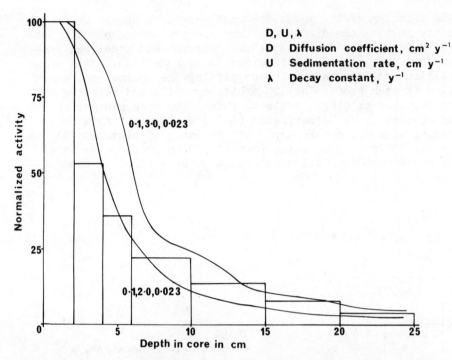

Fig. 2 Predicted distribution of Cs 137 for same core as fig 1 and the observed profile in histogram form.

fusion coefficient has been derived which gives the best agreement between the model predicted distribution and the observed core data. The diffusion coefficient so derived is 0.001 cm^{2}y^{-1}. Unfortunately no published evidence appears to exist of any direct measurements of the diffusion coefficient for plutonium having been made, either in the laboratory or the environment, with which to compare this value. It is, however, small and as such would appear consistent with the evidence of plutonium being an immobile element compared with others such as caesium. Furthermore, the model shows little sensitivity to changes in D between 0.001 and 0.1 cm^{2}y^{-1}, values over this range producing more or less the same profile. The contrasting strong dependence on sedimentation rate is consistent with the view that the rate of sedimentation of already contaminated material is the predominant factor governing the accumulation of plutonium in these estuarine deposits, post depositional modification of the profiles being undetectable.

The next step was to apply the model to the prediction of the caesium 137 profile in the above core. Sedimentation rates of 2 and 3 cm y^{-1} were used. The results are shown in Fig 2 for a range of values of the diffusion coefficient, D. Closest agreement is obtained with D = 0.1 cm^{2}y^{-1}.

Similar attempts have been made to match the caesium 137 profiles measured in two cores taken during the summer of 1980, one from the same area of the intertidal bank used for the earlier work on plutonium and the other from the edge of the stabilised marsh well colonised by plants and submerged only at high water spring tides. The intertidal core was one of three taken within an area of a square metre, all of which showed very similar activity profiles. With a diffusion coefficient of 0.1 cm^{2}y^{-1}, the model was run for sedimentation rates in the range 1 to 3 cm y^{-1}.

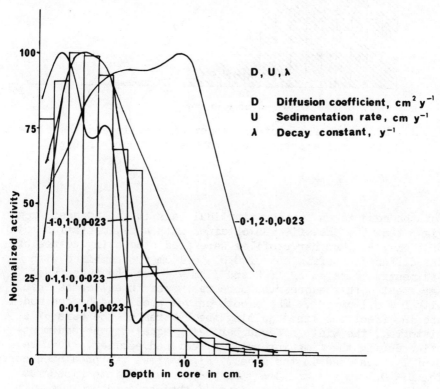

Fig 3. Predicted distribution of Cs 137 for core taken in 1980 from Newbiggin area of the Ravenglass Estuary and the observed profile in histogram form.

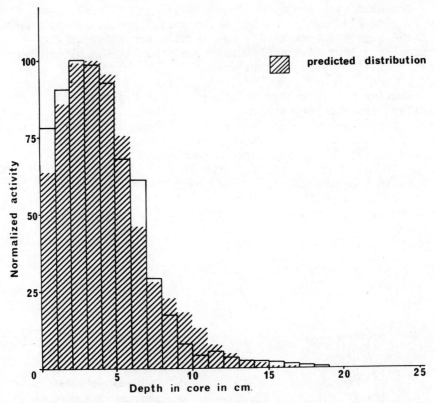

Fig 4. Histogram forms of the observed profile and the closest predicted distribution for the same core as fig 3.

For the core taken on the intertidal bank it can be seen from Fig 3 that the best fit was obtained with a sedimentation rate of 1 cm y^{-1}. Further profiles were predicted using values of the diffusion coefficient of 0.01 to 1 $cm^2 y^{-1}$ combined with sedimentation rates of 1, 2 and 3 cm y^{-1} but the measure of agreement with the observed core was worse in all cases than with $D = 0.1$ $cm^2 y^{-1}$. The smooth curves predicted by the model are in effect the lines joining the centres of the bars of a histogram, the width of each bar being equal to the sedimentation during the time interval, P, defined earlier. A value for P of 1 y was used in the model. Fig 4 shows the predicted curve for $D = 0.1$ $cm^2 y^{-1}$ and $U = 1$ cm y^{-1} converted into histogram form to permit easier comparison of the observed profile with the model predicted distribution giving the best fit of it.

It would appear therefore that the value of the diffusion coefficient for caesium 137 in these sediments is about $0.1 \text{ cm}^2\text{y}^{-1}$ which is in close agreement with the value published by Duursma and Bosch (1970) from laboratory studies on similar sediment types. The large difference between the true diffusion coefficient and the apparent coefficient derived previously with the model which took no account of sedimentation, emphasises again the importance of sedimentation in this area and its over-riding effect on the observed radioactivity profiles.

Fig 5 shows the activity profile of caesium 137 of a core taken from the seaward edge of stabilized marsh already colonised by vegetation and covered only by the highest tides. The fit is not as good as the other cores, a sedimentation rate of 2 cm y^{-1} being closest. A rate higher than that farther down the mud bank in the tidal area is to be expected. It might also be expected that the rate of sedimentation in this area could vary rather more from year to year than on the mud bank as high tides could well deposit quite variable amounts of material. It

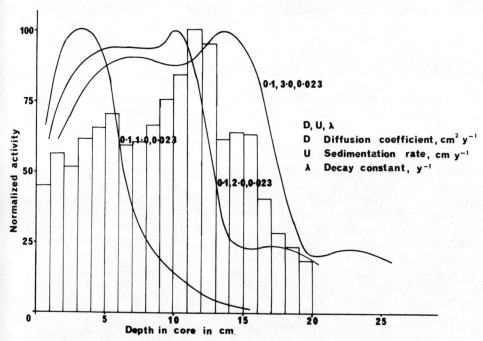

Fig.5. Observed histogram distribution of Cs 137 in core taken in 1980 from seaward edge of consolidated sea marsh at Newbiggin in Ravenglass Estuary and the predicted distribution.

is of interest to note the finer detail which can be seen not only in the profile predicted by the model but also in the core itself. The distribution observed with respect to depth follows the pattern with respect to time of caesium 137 discharges from Windscale. The preservation of such detail within the sediment is a further indication of the influence of sedimentation of contaminated material in the generation of the observed profiles and of the relatively minor importance of factors tending to rework the sediments once they have been deposited, such as biological activity and resuspension by wind or wave action. In the case of the area at the upper limit of the tidal range, where energies are approaching zero, resuspension by water action would probably not be anticipated and the successful establishment of vegetation tends to confirm that it is not happening to any significant degree. On the other hand, because sediments in such areas are finer grained and have more opportunity to dry out as compared with those washed by every tide, resuspension by the wind might be expected to occur more readily. Although such wind resuspension may occur from time to time (particularly it is believed of particulate material deposited on the vegetation itself) the present evidence suggests that it has little effect on the way in which the sediments consolidate.

APPLICATIONS FOR THE MODEL

Attention has been drawn before to the potential value of radionuclides in sediment studies generally (Hetherington, 1979). The present studies may be of particular interest in relation to the effects of biological activity in estuarine sediments in this area. Claims have been made from time to time that such activity has been responsible for modifying the depth profiles of radioactivity seen within the sediments in ways which envisage the wholesale reworking of the deposits and which would be expected to destroy completely the original sedimentary structure (Duursma and Gross, 1971). The earlier studies with plutonium were useful in showing that such reworking seemed very unlikely in view of the way in which depth profiles appeared to be preserved over many years (Hetherington, 1979). The present results give further confirmation of the conclusions reached from the plutonium work. All these applications are however essentially negative in that they produce no new information about the biological processes which must go on within the sediments, even though these processes do not appear to affect the depth distributions on a macroscopic scale. The next step would seem to be to refine the sampling techniques in an attempt to achieve a degree of resolution from the radionuclide work capable of permitting the direct observation of the effects of biological processes. In the longer term, such processes may be vitally important if they alter the chemical form of the pollutants so as to cause their remobilisation from the deeper sediments.

REFERENCES

Aston, S.R., and Stanners, D.A., 1979, Determination of estuarine sedimentation rates by caesium 134/caesium 137 and other artificial radionuclide profiles, Est. and Coastal Mar. Sci. 9:529.

Duursma, E.K., and Bosch, C.J. 1970, Theoretical, experimental and field studies concerning diffusion of radioisotopes in sediments and suspended solid particles of the sea, Part B Methods and Experiments, Neth. J. Sea Res. 4:395.

Duursma, E.K., and Gross, M.G., 1971, Marine sediments and radioactivity, in: "Radioactivity in the Marine Environment", National Academy of Sciences, National Research Council, Washington, D.C.

Hetherington, J.A., 1976, The bevaviour of plutonium nuclides in the Irish Sea, in: Environmental Toxicity of Aquatic Radionuclides - Models and Mechanisms", M.N. Miller and J.W. Stannard, eds., Ann Arbor Science Publishers, Ann Arbor Michigan.

Hetherington, J.A., 1979, The uptake of plutonium nuclides by marine sediments, Mar. Sci. Comm. 4:239.

Hetherington, J.A., and Jefferies, D.F., 1974, The distribution of some fission products in sea and estuarine sediments, Neth. J. Sea Res. 8:319.

Lerman, A., 1971, Transport of radionuclides in sediments, in: Proc. 3rd Nat. Symp. on Radioecology, NTIS, Springfield, Illinois.

The Windscale Inquiry, 1978, HMSO, London.

DISTRIBUTION, COMPOSITION AND SOURCES OF POLYCYCLIC AROMATIC HYDROCARBONS IN SEDIMENTS OF THE RIVER TAMAR CATCHMENT AND ESTUARY, U.K.

J.W. Readman[1], R.F.C. Mantoura[2], and M.M. Rhead[1]

[1] Department of Environmental Sciences,
Plymouth Polytechnic, Drake Circus, Plymouth, Devon, UK

[2] Institute for Marine Environmental Research,
Prospect Place, The Hoe, Plymouth, Devon, UK

INTRODUCTION

The polycyclic aromatic hydrocarbons (PAH) considered in this study are a family of 3-6 ringed aromatic compounds whose structures are shown in Figure 1.

The carcinogenic properties of some PAH coupled with their ubiquity in the environment have, in recent years, led to interest in their sources, distribution, transport mechanisms and fate. Many reviews are available on these topics: (Andelman and Snodgrass, 1972; Andelman and Suess, 1970; Bjørseth and Dennis, 1980; Brown, 1979; Harrison et al., 1975; Jackim and Lake, 1978; Jones and Leber, 1979; Neff, 1979, and Radding et al., 1976).

Sedimentary PAH have received considerable attention because they represent a record of the input history of the compounds into an aquatic environment. The occurrence and distribution of PAH in surface sediments have been extensively investigated: (Bieri et al., 1978; Blumer and Youngblood, 1975; Giger and Schaffner, 1978; John et al., 1979; Laflamme and Hites, 1978; Lake et al., 1979; Matsushima, 1979; Thompson and Eglinton, 1978 and others). More recently, the history of PAH inputs have been assessed by analysing PAH distribution in dated sediment cores (Grimmer and Bohnke, 1975; Heit et al., 1981; Hites et al., 1977; Hites et al., 1980; Muller et al., 1977; Platt and Mackie, 1979; Wakeham et al., 1980). In many of these papers attempts have been made to ascertain the sources of the PAH by comparing the homologue compositions within the sedi-

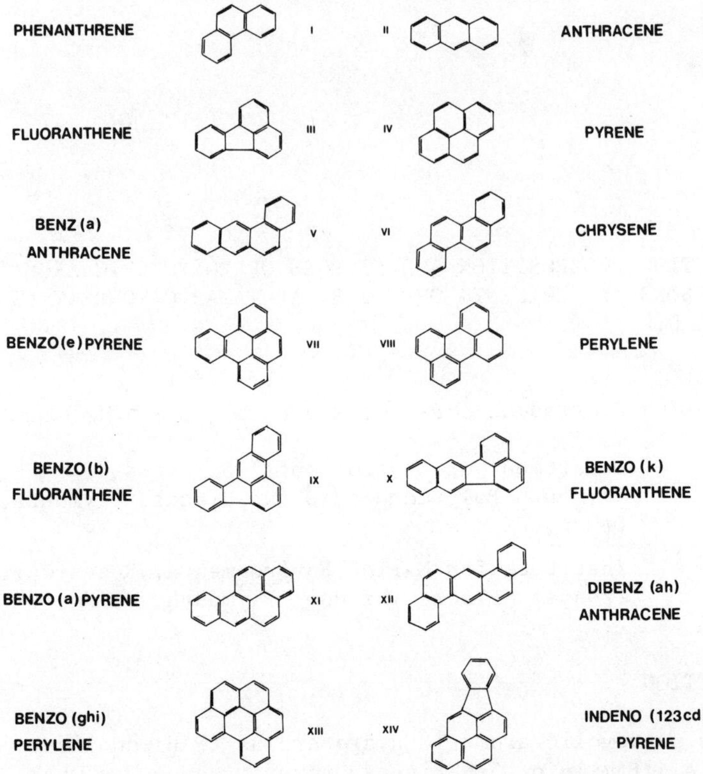

PHENANTHRENE   I   II   ANTHRACENE

FLUORANTHENE   III   IV   PYRENE

BENZ(a) ANTHRACENE   V   VI   CHRYSENE

BENZO(e)PYRENE   VII   VIII   PERYLENE

BENZO(b) FLUORANTHENE   IX   X   BENZO(k) FLUORANTHENE

BENZO(a)PYRENE   XI   XII   DIBENZ(ah) ANTHRACENE

BENZO(ghi) PERYLENE   XIII   XIV   INDENO(123cd) PYRENE

Fig. 1. Structures of PAH considered in this study. The Roman numerals adjacent to the individual compounds are used in the text to draw attention to the molecular structures.

mentary PAH assemblages with known pollutant emissions. Proposed primary sources include combustion of fossil fuels, (Heit et al., 1981; Hites et al., 1977; Hites et al., 1980; Laflamme and Hites, 1978), combustion primarily of coal (Grimmer and Bohnke, 1975; Muller et al., 1977), road runoff/street dust (Giger and Schaffner, 1978; Wakeham et al., 1980), the liberation of 'natural' PAH by erosion of coal (John et al., 1979) and natural fires (Blumer and Youngblood, 1975). Other sources deemed of more localised significance but worthy of note include oils, domestic (sewage) and industrial waste water, creosote and tars. PAH from biosynthesis has also been proposed but remains inconclusively demonstrated (Neff, 1979).

PAH are distributed over large distances via transport of airborne combustion particulates (Hites et al., 1978; Laflamme and Hites, 1978, and others). Within localised areas, however, information on transport is less readily available. Hites et al. (1978)

propose a theory of resuspension and transport of contaminated sediments away from urban areas. This is supported by the geographical distribution of PAH in sediments of Massachusetts Bay and the Gulf of Maine.

The distribution and behaviour of PAH transported from fresh water into the marine environment involves the passage of PAH through steep biological, chemical and hydrodynamic gradients. These factors, often coupled with the proximity of large urban and industrial conurbations, render estuaries of particular interest, though detailed information on PAH is sparse.

This paper describes the distribution of selected PAH in sediments of the River Tamar catchment and estuary. Major sources of the compounds were investigated and transport mechanisms that could produce the observed distribution are evaluated.

STUDY AREA AND SAMPLE COLLECTION

The geographical location of the sampling area and position of the sample sites are illustrated in Figure 2. The Tamar catchment drains relatively unpolluted moorland. In contrast, urban Plymouth is situated adjacent to the lower estuary (Figure 2). Morris et al. (1982) have described the master variable chemistry of the Tamar Estuary and the aquatic distribution and microbial heterotrophic degradation of PAH have been reported by Readman et al. (1982). Although the Lynher and Tavy tributaries enter the estuary, the Tamar represents the major fresh-water source, consequently sampling was restricted to this river along the main estuarine axis.

River sediment samples were removed using a glass corer (5 cm diameter) to a maximum depth of 5 cm. Estuarine sediment core samples (2.5 cm diameter, 10-15 cm depth) were obtained using the Butler gravity corer (Butler and Tibbits, 1972), the top 5 cm sections being retained for analysis. Sediment samples were placed in acid-cleaned jars sealed with solvent-rinsed aluminium foil liners, which were immediately stored over solid carbon dioxide, then maintained frozen in the laboratory until extraction.

METHODS

Sub-samples of the thawed homogenised sediments were freeze dried to determine water content. Catchment sediments were then sieved to evaluate particle size distribution and the estuarine muds were ground in an agate ball mill and then analysed for carbon content (Carlo Erba Elemental Analyser Model 1106).

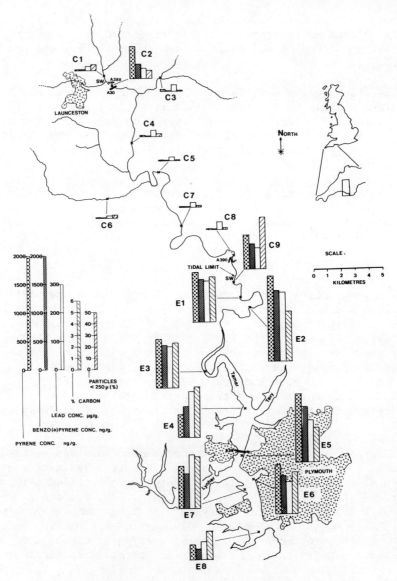

Fig. 2. Description of the Tamar catchment and estuary showing sediment sample site locations (C1-C9 and E1-E8). Pyrene (IV), benzo(a)pyrene (XI) and lead concentrations and the percentage of particles sized <250 μm and % carbon content data are shown graphically for each sample. Scales for the graphs are indicated on the left hand side of the diagram. Urban regions in the area are demarked by open stipples. Roads crossing the Tamar are shown and denoted by their numbers. Sewage works of particular note are indicated by 'SW'.

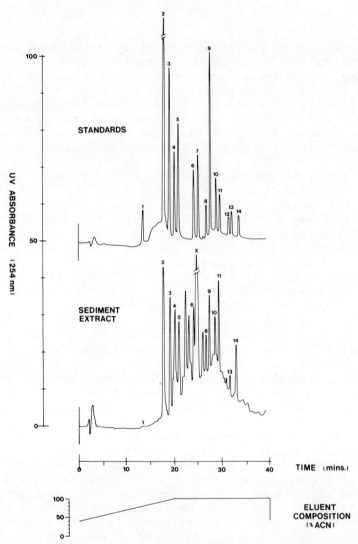

Fig. 3. Typical HPLC chromatograms showing PAH separations of standards and a sediment extract (C2) using a Perkin Elmer 10 μm HC ODS (250 x 26 mm i.d.) column. Peak numbers correspond to 1. naphthalene, 2. phenanthrene (I), 3. anthracene (II), 4. fluoranthene (III), 5. pyrene (IV), 6. benz(a)anthracene (V), 7. chrysene (VI), 8. benzo(e)-pyrene (VII), 9. coincident perylene/benzo(b)fluoranthene (VIII/IX), 10. benzo(k)fluoranthene (X), 11. benzo(a)-pyrene (XI), 12. dibenz(ah)anthracene (XII), 13. benzo(ghi)-perylene (XIII), 14. indeno(1,2,3-cd)pyrene (XIV). X is an unresolved mixture (see text). Elution conditions are shown at the bottom of the figure. Detection was by UV absorbance at 254 nm.

PAH Analysis

Full details of the extraction, clean-up and chromatographic procedures utilised are described elsewhere (Readman et al., 1982). In summary, wet sediments were ground with ashed anhydrous sodium sulphate and then subjected to soxhlet extraction with dichloromethane (8 hours in darkness). Portions of the resulting extracts were concentrated and cleaned-up by passage through alumina. The resulting hexane fraction was displaced with acetonitrile and the sample stored in darkness at $0^\circ C$ awaiting analysis.

The estuarine samples were analysed by high-performance liquid chromatography (HPLC) according to Readman et al. (1982) using a Hypersil 5.7 μm (250 x 50 mm i.d.) octadecylsilane (ODS) column with variable wavelength UV detection. This system was modified for the catchment samples to increase resolution by the use of a Perkin-Elmer 10 μm HC-ODS (250 x 26 mm i.d.) PAH specific column fitted with a Whatman pellicular ODS guard column. The chromatographic method utilised was similar to that described by Ogan et al. (1979). An acetonitrile: distilled water elution system was chosen at a flow rate of 0.5 cm^3 min^{-1}. The solvent programme consisted of an isocratic stage (40% acetonitrile: 60% distilled water) for 15 minutes prior to injection and on injection a gradient increase of 3% acetonitrile. min^{-1} was introduced. Once the solvent composition of 99% acetonitrile: 1% distilled water was reached this concentration was maintained until completion of the chromatogram. UV absorbance detection of the compounds was monitored at 254 nm and 280 nm. Initial peak identification was performed by coinjection with standard PAH. Ratios of absorbance at 254:280 nm were calculated for all sample peaks for comparison with those of standards. Additional information on identification/peak purity was obtained by comparison of stop-flow UV scans of environmental sample separations with those of authentic standards (Readman et al., 1981). Calibration of the system was achieved using known concentrations of high purity PAH in acetonitrile. Examples of chromatograms obtained are shown in Figure 3. The large peak (X) in the environmental sample (Figure 3) with a retention time of 25 minutes, exhibited a UV spectrum comparable to that obtained for the peak identified from the Hypersil separation (Readman et al., 1982) as a mixture of phenyl esters and phthalates. In addition, chrysene (VI) coelutes with this peak.

Lead Analysis

Samples were digested with HCl - HNO_3 according to the method of Van Loon et al., (1973) and analysed using an Instrumentation Laboratory Incorporated atomic absorption spectrophotometer.

RESULTS AND DISCUSSION

Concentrations

Concentrations of individual PAH in the Tamar estuary sediments (0-5 cm) varied typically between 30 to 1500 ng(g dry sediment)$^{-1}$. These values are comparable with those of 100 to 5200 ng(g dry sediment)$^{-1}$ reported for Severn Estuary (U.K.) sediments, (John et al., 1979; Thompson and Eglinton, 1978) and for estuarine sediments from Narragansett Bay, U.S.A. (Lake et al., 1979). Tamar river sediments, however, contained considerably lower concentrations ranging typically from less than 1 to 50 ng(g dry sediment)$^{-1}$, levels which are indicative of a relatively unpolluted environment (Neff, 1979).

Distribution

Figure 2 shows the spatial distribution of pyrene (IV) and benzo(a)pyrene (XI) as representative PAH throughout the sampling area. The riverine sediments are relatively depleted of PAH and lead, which is consistent with the generally unpolluted state of the catchment. In addition, any small particulates with their associated PAH and light organic material onto which PAH would preferentially adsorb (Herbes, 1977; Karickhoff et al., 1979; Means et al., 1980) are likely to be selectively removed by the fast flow of the river. The only exception to this trend of low PAH concentrations in fresh water sediments is sample C2 which is exposed to chronic anthropogenic pollution. The PAH levels at station C4 located 5 Km downstream of C2 however, return to the low concentrations typical of the catchment. The influence of the PAH source on sediment concentrations is therefore relatively short lived.

In Figure 2 the percentage weights of particles <250 µm are indicated for the catchment survey sediments (composed of a sand/gravel mixture) but are replaced by carbon analyses for the estuarine mud samples. There is a hundred fold increase in the concentrations of PAH from C8 to C9, with high concentrations maintained throughout the estuary. This surge in the PAH concentrations occurs just downstream from the weir which corresponds to the tidal limit of the estuary. Samples C9, E1 and E2 contain substantially elevated PAH levels which increase sequentially. Settlement and flocculation processes in this intertidal region depositing PAH rich riverine particulates are the mechanisms most likely to produce this observed distribution. This is consistent with the substantial increase in the proportion of particles <250 µm at station C9. Maher et al. (1979) report a greater than 10 fold increase in concentrations of benzo(a)pyrene from fresh water sediments to estuarine sediments in the River Yarra, South East Australia. Further towards the mouth of the estuary at stations E3 and E4, PAH content decreases probably owing to dilution with less polluted marine sediments together with

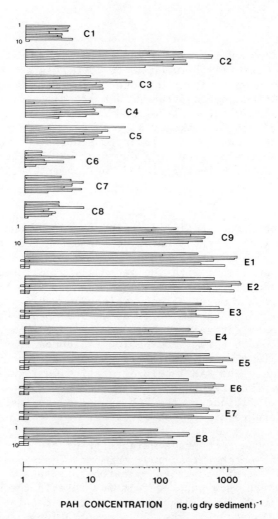

Fig. 4. Compositional variations of PAH in Tamar sediments. For each sample 10 lines are shown indicating individual PAH concentrations starting with 1. phenanthrene (I) at the top and proceeding downwards in sequence - 2. anthracene (II), 3. fluoranthene (III), 4. pyrene (IV), 5. benz(a)anthracene (V), 6. benzo(e)pyrene (VII), 7. benzo(k)fluoranthene (X), 8. benzo(a)pyrene (XI), 9. benzo(ghi)perylene (XIII), 10. indeno(1,2,3-cd)pyrene (XIV). To aid comparison between samples a log. PAH concentration scale was selected. Short lines extending beyond the base line indicate that these compounds were not quantified.

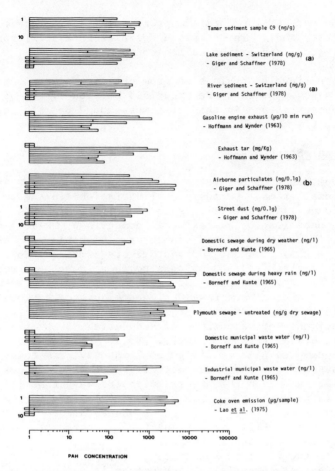

Fig. 5. PAH compositions discussed in the 'Sources' section of this report. For comparison purposes Tamar sediment sample C9 is included. For each assemblage 10 lines are shown indicating individual PAH concentrations starting with 1. phenanthrene (I) at the top and proceeding downwards in sequence – 2. anthracene (II), 3. fluoranthene (III), 4. pyrene (IV), 5. benz(a)anthracene (V), 6. benzo(e)pyrene (VII), 7. benzo(k) fluoranthene (X), 8. benzo(a)pyrene (XI), 9. benzo(ghi) perylene (XIII), 10. indeno(1,2,3-cd)pyrene (XIV). To aid comparison between samples a log. PAH concentration scale was selected. Units of concentration vary and are individually noted. Short lines extending beyond the base line indicate that these compounds were not quantified.

a For these samples further qualitative information of compounds not quantified is available from chromatograms published by the authors.

b The authors suggest that the sampling procedure is responsible for loss of the lower molecular weight homologues.

degradation and physico/chemical processes (see 'Compositional Structure' section). This seaward decrease is also mirrored by an overall trend of decreasing carbon content.

Elevated levels originating from chronic anthropogenic sources are again associated with the urbanised portion of the estuary. The highest concentrations in this area were found beneath the A38 Tamar Road Bridge (Figure 2).

Compositional Structure

The dominance of unsubstituted parent hydrocarbons was apparent in all chromatograms. Compositional variations in the suite of compounds quantified are illustrated in Figure 4. These are plotted on a log. concentration scale to aid comparison. PAH composition of the estuarine sediments E1-E8 and the two high PAH concentration samples from the catchment (C2 and C9) are similar. Compositionally similar sedimentary PAH assemblages to these have been reported for the Severn Estuary (Thompson and Eglinton, 1978), in river sediment from Monchaltorfer Aa and Lake Greifensee surface sediment, Switzerland (Giger and Schaffner, 1978) (see Figure 5) and in surface sediments of Lakes Lucerne and Zurich, Switzerland and Lake Washington, U.S.A. (Wakeham et al., 1980). Differences are however noted with samples E1 to E4 (Figures 2 and 4). PAH concentrations decrease seaward but the lower molecular weight compounds decrease at a greater rate than the higher molecular weight homologues. Sample E4 is thus enriched with eg. benzo(a)pyrene (XI) or, alternatively, relatively depleted of phenanthrene (I), anthracene (II), fluoranthene (III) and pyrene (IV). Degradation and/or solubilization are probably responsible for the preferential reduction in the relative abundance of the lower molecular weight PAH. The general compositional structure is restored in the urban portion of the estuary. This highlights the necessity for consideration of _in situ_ changes in composition when considering sources. Catchment samples C1 and C3 - C8 exhibit differences from this general trend to varying degrees, C1 appearing particularly anomalous.

Sources

For comparison purposes, the compositional PAH assemblages associated with various source emissions reported in the literature are illustrated in Figure 5. Included are: motor vehicle exhaust, airborne particulates, street dust, sewage, domestic and industrial waste water and coke oven emissions, and together these are likely to be representative of the major sources of PAH in the Tamar area. The dominance of unsubstituted parent hydrocarbons in the chromatograms is generally indicative of combusion sources and exclude the more complex assemblages associated with undegraded or native oils.

It is also of note that the PAH compositional patterns at the riverine (C2) and lower estuarine sampling stations are similar, tending to suggest similar primary inputs.

As discussed in the 'Compositional Structure' section of this report, other researchers eg. Giger and Schaffner (1978)(Figure 5), and Wakeham et al. (1980), have recorded similar PAH assemblages to those observed in the Tamar. These authors attribute 'street dust' as the major source. Giger and Schaffner (1978) also suggest a similarity of composition of airborne particulates (Figure 5) and Wakeham et al., (1980) emphasise the importance of asphalt particles in street runoff. Sample stations C2 and C9 are located in close proximity to major roads and are likely to receive polluted runoff. In addition, sample E5, situated below the Tamar road and rail bridges showed the highest PAH concentrations of the lower estuary. The pattern of lead distribution, a possible indicator of motor vehicle traffic (Greenberg et al., 1981) is shown in Figure 2. Lead plotted against benzo(a)pyrene (taken as an exemplary PAH) (Figure 6) shows a near linear relationship (correlation coefficient = 0.90) suggesting that both pollutants might originate from the same source. Recent studies of a dated sediment core (in preparation) taken from the lower estuary show dramatic increases in PAH and lead concentrations since the late 1940's. From the combined evidence, it is likely that road runoff is a primary source of PAH into the Tamar. The exhaust emission shown in Figure 5 is clearly not entirely responsible for the road runoff composition. It is probable that a combination of exhaust emissions, asphalt particles, sump oil and tyre wear are involved.

Other potential sources investigated included sewage. Sample sites C2, C9 and E6 are close to sewage works and the overall influence of sewage on the urban portion of the estuary (as indicated by increases in sediment carbon contents - Figure 2) rendered these emissions of interest. PAH distribution in the lower estuary, however, appears axially more confined than the sewage influenced area (Readman et al., 1982). Analysis of raw sewage from Plymouth resulted in a more complex chromatogram enriched with lower molecular weight compounds (Figure 5) and alkylated homologues. As can be seen from other sewage analyses (Figure 5) considerable variability of composition can be expected dependent on the degree of domestic and industrial waste and the extent to which street runoff contributes to storm overflows. An important proportion of PAH from the sewage works is likely to originate from road runoff during heavy rain (Borneff and Kunte, 1965).

The PAH assemblage from coke oven emission (Figure 5) resembles, to an extent, the sedimentary composition. Airborne combustion particulates from domestic coal burning in the area are unlikely to

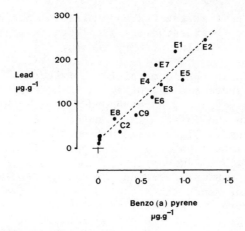

Fig. 6. Relationship between lead and benzo(a)-pyrene concentrations ($\mu g.(g$ dry sediment$)^{-1}$) in Tamar surface sediments. Samples containing high concentrations are individually identified (positions of sampling are shown in Figure 2). Linear regression of the individual points is indicated by the broken line (correlation coefficient = 0.90).

result in the observed sediment core and estuarine spatial distribution of the compounds, precluding these as a primary source. It is also unlikely that the prevailing south westerly air stream in the region, which originates from the Atlantic Ocean, would contain high concentrations of PAH.

The distribution of coal particles at the top of the estuary have qualitatively been linked to the dilution described for PAH. However, the compositional patterns of PAH derived from coal do not resemble those observed in this study (John et al., 1979; Tripp et al., 1981).

An important consideration when assessing sources by direct comparison with the parent material are homologue selective changes that could occur after emission of PAH into the environment. We have discussed changes observed in sedimentary PAH composition in the upper region of the estuary where lower molecular weight PAH appear to be preferentially 'diluted'. This is identified as an in situ change rather than different source inputs primarily by the gradation involved. Owing to the fact that the urban sedimentary PAH composition closely resembles that at the top of the estuary and that recorded for polluted catchment sample C2 (Figure 2), it seems likely that a primary source dominates the Tamar area, and from the evidence discussed it would appear that the primary source is road

runoff. The absorbing capacity of sediments together with their
dynamic nature in the estuary might effectively integrate secondary
inputs, particularly in the urban region of the estuary.

CONCLUSIONS

Concentrations of individual PAH in sediments from the Tamar
catchment varied typically between less than 1 to 50 ng(g dry
sediment)$^{-1}$ whereas those from the estuary were between 30 to 1500 ng
(g dry sediment)$^{-1}$. A point source PAH emission in the river was
observed but its downstream distribution declined over a short
distance (<5 Km).

The generally low levels of PAH in the river are attributed to
the relatively pristine environment and the fast flow removing PAH
rich particulates. Sedimentation/flocculation processes at the head
of the estuary deposit riverine particulates, substantially increasing the sedimentary PAH concentrations. Levels then decrease
towards the mid-estuarine region probably owing to dilution with less
polluted marine sediments together with degradation and physico/
chemical removal processes. Further elevated levels associated with
the urbanised portion of the estuary originate from chronic
anthropogenic sources.

PAH compositions of the estuarine sediments E1-E8 and the two
high PAH concentration samples from the catchment (C2 and C9) are
similar.

The PAH assemblage and covariability with lead, together with
distribution of PAH in a dated sediment core all suggest that road
runoff is a primary source of PAH to the Tamar.

ACKNOWLEDGEMENTS

The authors wish to thank Dr. L. Brown of Plymouth Polytechnic
for assistance and advice.

Dr. Mantoura's role in this research forms a part of the
estuarine research program of the Natural Environmental Research
Council, and was partly supported by the Department of the
Environment (contract DGR480/48).

REFERENCES

Andelman, J.B., and Snodgrass, J.E., 1972, Incidence and significance
of polynuclear aromatic hydrocarbons in the water environment,
CRC Crit. Rev. Environ. Contr., 4: 69.

Andelman, J.B., and Suess, M.J., 1970, Polynuclear aromatic hydrocarbons in the water environment, Bull. Wld. Hlth. Org., 43: 479.

Bieri, R.H., Kent-Cueman, M., Smith, C.L., and Chih-Wu Su, 1978, Polynuclear aromatic and polycyclic aliphatic hydrocarbons in sediments from the Atlantic Outer Continental Shelf, Intern. J. Environ. Anal. Chem., 5: 293.

Bjørseth, A., and Dennis, A.J., 1980, "Polynuclear aromatic hydrocarbons: chemistry and biological effects", Proceedings of the Fourth International Symposium on PAH, Batelle Press, Columbus.

Blumer, M., and Youngblood, W.W., 1975, Polycyclic aromatic hydrocarbons in soils and Recent sediments, Science, 188: 53.

Borneff, J., and Kunte, H., 1965, Carcinogenic substances in water and soil. Part XVII: Concerning the origin and estimation of the polycyclic aromatic hydrocarbons in water, Arch. Hyg. (Berlin), 149: 226, (German).

Brown, R.A., 1979, "Fate and effects of polynuclear aromatic hydrocarbons in the aquatic environment", American Petroleum Institute. Environmental Affairs Dept., Publication Number 4297.

Butler, E.I., and Tibbits, S., 1972, Chemical survey of the Tamar Estuary, J. Mar. Biol. Assn. U.K., 52: 681.

Giger, W., and Schaffner, C., 1978, Determination of polycyclic aromatic hydrocarbons in the environment by glass capillary gas chromatography, Anal. Chem., 50: 243.

Greenberg, A., Bozzelli, J.W., Cannova, F., Forstner, E., Giorgio, P., Stout, D., and Yokoyama, R., 1981, Correlations between lead and coronene concentrations at urban, suburban and industrial sites in New Jersey, Environ. Sci. Technol., 15: 566.

Grimmer, G., and Bohnke, H., 1975, Profile analysis of polycyclic aromatic hydrocarbons and metal content in sediment layers of a lake, Cancer Lett., 1: 75.

Harrison, R.M., Perry, R., and Wellings, R.A., 1975, Polynuclear aromatic hydrocarbons in raw, potable and waste waters, Water Res., 9: 331.

Heit, M., Tan, Y., Klusek, C., and Burke, J.C., 1981, Anthropogenic trace elements and polycyclic aromatic hydrocarbon levels in sediment cores from two lakes in the Adirondack acid lake region. Water, Air and Soil Pollut., 15: 441.

Herbes, S.E., 1977, Partitioning of polycyclic aromatic hydrocarbons between dissolved and particulate phases in natural waters, Water Res., 11: 493.

Hites, R.A., Laflamme, R.E., and Farrington, J.W., 1977, Sedimentary polycyclic aromatic hydrocarbons: The historical record, Science, 198: 829.

Hites, R.A., Laflamme, R.E., and Windsor, J.G., 1978, Polycyclic aromatic hydrocarbons in marine/aquatic sediments, in: "Symposium on analytical chemistry of petroleum hydrocarbons in the marine/aquatic environment, Miami Beach meeting, September 10-15, 1978", American Chemical Society.

Hites, R.A., Laflamme, R.E., Windsor, J.G., Farrington, J.W., and Deusser, W.G., 1980, Polycyclic aromatic hydrocarbons in an anoxic sediment core from the Pettaquamscutt River (Rhode Island, U.S.A.), Geochim. Cosmochim. Acta, 44: 873.

Hoffmann, D., and Wynder, E.L., 1963, Studies on gasoline engine exhaust, J. Air Pollut. Contr. Assn., 13: 322.

Jackim, E., and Lake, C., 1978, Polynuclear aromatic hydrocarbons in estuarine and nearshore environments, in: "Proceedings of the Fourth Biannual International Estuarine Research Conference, Pennsylvania, October 2-5 (1977)", M.L. Wiley (ed.), Academic Press, London.

John, E.D., Cooke, M., and Nickless, G., 1979, Polycyclic aromatic hydrocarbons in sediments taken from the Severn Estuary drainage system, Bull. Environm. Contam. Toxicol., 22: 653.

Jones, P.W., and Leber, P., 1979, "Polynuclear Aromatic Hydrocarbons", Third International Symposium on Chemistry and Biology, Carcinogenesis and Mutagenesis, Ann Arbor Science, Ann Arbor.

Karickhoff, S.W., Brown, D.S., and Scott, T.A., 1979, Sorption of hydrophobic pollutants on natural sediments, Water Res., 13: 241.

Laflamme, R.E., and Hites, R.A., 1978, The global distribution of polycyclic aromatic hydrocarbons in Recent sediments, Geochim. Cosmochim. Acta, 42: 289.

Lake, J.L., Norwood, C., Dimock, C., and Bowen, R., 1979, Origins of polycyclic aromatic hydrocarbons in estuarine sediments, Geochim. Cosmochim. Acta, 43: 1847.

Lao, R.C., Thomas, R.S., and Monkman, J.L., 1975, Computerized gas chromatographic-mass spectrometric analysis of polycyclic aromatic hydrocarbons in environmental samples, J. Chromatog., 112: 681.

Matsushima, H., 1979, Correlation of polynuclear aromatic hydrocarbons with environmental components in sediment from Hirakata Bay, Japan, Agric. Biol. Chem., 43: 1447.

Maher, W.A., Bagg, J., and Smith, J.D., 1979, Determination of polycyclic aromatic hydrocarbons in marine sediments, using solvent extraction, thin-layer chromatography and spectrofluorimetry, Intern. J. Environ. Anal. Chem., 7: 1.

Means, J.C., Wood, S.G., Hassett, J.J., and Banwart, W.L., 1980, Sorption of polynuclear aromatic hydrocarbons by sediments and soils, Environ. Sci. Technol., 14: 1524.

Morris, A.W., Bale, A.J., and Howland, R.J.M., 1982, Chemical variability in the Tamar Estuary S.W. England, Estuarine Coastal Shelf Sci., 14: 649.

Muller, G., Grimmer, G., and Bohnke, H., 1977, Sedimentary record of heavy metals and polycyclic aromatic hydrocarbons in Lake Constance, Naturwissen., 64: 427.

Neff, J.M., 1979, "Polycyclic aromatic hydrocarbons in the aquatic environment – sources, fates and biological effects", Applied Science, London.

Ogan, K., Katz, E., and Slavin, W., 1979, Determination of polycyclic aromatic hydrocarbons in aqueous samples by reversed-phase liquid chromatography, Anal. Chem., 51: 1315.

Platt, H.M., and Mackie, P.R., 1979, Analysis of aliphatic and aromatic hydrocarbons in Antarctic marine sediment layers, Nature, 280: 576.

Radding, S.B., Mill, T., Gould, C.W., Liu, D.H., Johnson, H.L. Bomberger, D.C., and Fojo, C.V., 1976, "The environmental fate of selected polynuclear aromatic hydrocarbons". EPA560/5-75-009, U.S. Environmental Protection Agency, Washington, D.C.

Readman, J.W., Brown, L., and Rhead, M.M., 1981, Use of stop-flow ultraviolet scanning and variable-wavelength detection for enhanced peak identification and sensitivity in high-performance liquid chromatography, Analyst, 106: 122.

Readman, J.W., Mantoura, R.F.C., Rhead, M.M., and Brown, L., 1982, Aquatic distribution and heterotrophic degradation of polycyclic aromatic hydrocarbons (PAH) in the Tamar Estuary, Estuarine Coastal Shelf Sci., 14: 369.

Thompson, S., and Eglinton, G., 1978, Composition and sources of pollutant hydrocarbons in the Severn Estuary, Mar. Pollut. Bull., 9: 133.

Tripp, B.W., Farrington, J.W., and Teal, J.M., 1981, Unburned coal as a source of hydrocarbons in surface sediments, Mar. Pollut. Bull., 12: 122.

Van Loon, J.C., Lichwa, J., Ruttan, D., and Kinrade, J., 1973, The determination of heavy metals in domestic sewage treatment plant works, Water, Air and Soil Pollut., 2: 473.

Wakeham, S.G., Schaffner, C., and Giger, W., 1980, Polycyclic aromatic hydrocarbons in Recent lake sediments – 1. Compounds having anthropogenic origins, Geochim. Cosmochim. Acta, 44: 403.

WATER QUALITY ASPECTS OF DUMPING DREDGED SILT INTO A LAKE

M. Veltman

Municipality of Rotterdam, Pb 6633 Rotterdam
The Netherlands

INTRODUCTION

The size of ships entering Rotterdam harbour is increasing all the time and extra dredging of the channels giving access to the ports is required. The real problem is to get rid of the dredged spoil. The quantities are enormous: 8×10^6 cubic metres of mud are dredged from the Rotterdam harbour-basins every year.

In the past there were no disposal problems at all. In the vicinity of Rotterdam there is plenty of low-lying land, protected by dykes from the higher water levels in the rivers. As this land was very wet, windmills were built to pump the water out and as a result the dewatered land consolidated and became even lower. Moreover, the soil was rich in peat which was a much sought household fuel in the old days. The lucrative peat digging lowered the ground level still further so that polder land four or five metres below sea level became quite common in the western Netherlands.

The chance to improve this situation by raising the land with material dredged up from the Rotterdam ports was welcomed by all concerned. After dumping the dredged spoil on the low-lying land the water in the mud will evaporate, the mud will break up and air will penetrate and many physical, chemical and biological reactions will occur, which result in a change in the structure of the mud. The mud will "ripen" to a good soil on which vegetation will thrive, and apart from the fertility, the higher land will give fewer dewatering problems. By "making work gives birth to work", as the Dutch say, two aims could be served:

- the ports would be deepened and made accessible to larger ships,

- the recycling of the dredged material would make low-lying land suitable to farming.

However the polluting effects of waste chemicals in the dredged material have become apparent. Getting rid of waste products is a problem which industry used to solve by discharging them with waste water. In the river the waste chemicals become attached to the silt which is carried out to sea or is deposited in the quieter waters of the harbour of Rotterdam.

Disposal problems arise as pollution grows worse, as shown by analyses of the Institute for Soil Fertility in 1972. In the 1960s disposal of the sediment dredged from the ports was less of a problem. At the moment, however, finding suitable disposal areas is one of the port authorities' chief headaches. Political pressure is still growing as people become aware of the environmental problems and as a result the idea of disposing of dredged silt in pits or lakes was born. An important question, however, concerns the effect of this material on the water quality of the lake. Disposal of dredged material in deep pits in port basins or river-beds does not seem to cause any special problems as long as such pits are in areas where there is a circulation of materials similar to the dredged material. However, when "dredge spoil" is deposited in pits or lakes where no contaminated sediments are present, problems crop up. Then we are confronted with questions such as:-

- What is the effect of turbidity?
- What is the effect of dumping on dissolved oxygen levels? (For this the organic content of the added material is important).
- What happens to organisms when they are exposed to pollutants and digest them?

As the lakes in the Rotterdam region are used for recreation, and deposits must not interfere with this use, it was proposed to leave a depth of, say, five metres of water. An important question to answer concerned the changes which might occur during and after dumping. To investigate this, an experimental site was selected. After an account of the purpose of the research programme this paper describes the experiments and presents some of the results.

To investigate the water quality aspects of the disposal of dredged material in the lake, a number of calculations were carried out which are described. Finally, possible measures to improve the water quality in the lake are examined.

RESEARCH PROGRAMME

The port department together with the State Waterway Board, which is responsible for the administration of rivers and Dutch territorial waters, commissioned the Hydraulic Laboratory (Delft) and the Institute for Soil Fertility (Groningen) to investigate the quality of water which might be expected in a lake into which dredge spoil had been dumped. The purpose of this programme was to determine whether certain variables such as nutrients, bacteria and contaminants would be above or below critical levels.

Points of Departure

As a case study, the Oostvoornse Meer had been selected (Fig. 1). This lake found its origin in the damming of the river Het Brielse Gat. In order to make provision for the expansion of industry a new area, called the Maasmond, was reclaimed in front of the Dutch coast. This was done by dredging sand out of the Oostvoornse Meer and that is why, in places, the lake has a depth of 40 m. Its average depth is 21·50 m. The total volume of the lake is 47 million m^3 and its surface area is 2·75 Km^2 and it is salt water. The bottom and the slopes consist mainly of sand, although silt occurs in deep pits on the bottom.

The lake was chosen following advice from the Municipality of Rotterdam in 1975, Delft Hydraulic Laboratory in 1976, Projectgroep Brielse Gat in 1978 and Swart, C.J. in 1977. The next important question is to ask what kind of dredged silt should be used for dumping? Here we think of what quality and density of silt to choose. We considered this to be so important that we chose three different rates of contamination and four different densities for study (Delft Hydraulic Laboratory Report 1979).

The next item to investigate, is the influence of time; in other words how long will it take to fill the lake and this largely depends on the quantity to be dumped annually. It is also evident that during summer you cannot dump into a lake used for recreation, so we decided to dump only in winter, from October to May inclusive.

Aspects

Having chosen the lake, the fill and the time, the physical and chemical processes that are environmentally important (see Fig. 2) are the next things to be examined.

- Firstly we have mixing, during dredging, during transport, and during dumping into the lake.

- Secondly we have sedimentation and erosion. The amount of sediment particles in the lake is very important, not only because

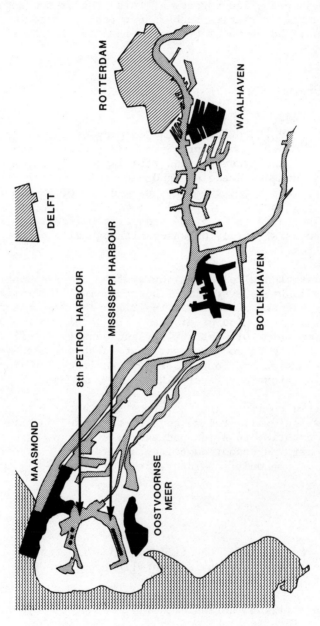

Fig. 1. Location of various places named in the text and the site of the experiment at Oostvoornse Meer.

WATER QUALITY ASPECTS OF DUMPING DREDGED SILT

HARBOUR

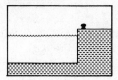

quality surface water
quality silt
quality pore water

DREDGING & TRANSPORT

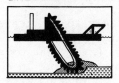

mixing
chemical reactions

DUMPING

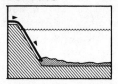

mixing
whirling up
sedimentation
chemical reactions

AFTERWARDS

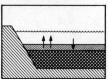

erosion
settling
flux of pore water
chemical reactions in pore water
chemical reactions in surface water

Fig. 2. Factors influencing water quality at a dredged material disposal site.

turbidity is a very important parameter for water quality, but also because most of the contaminants are attached to the particles.

- Thirdly, the process that influences the water quality most is the consolidation of the dredged material so that interstitial pore-water is pressed out. This pore-water has high concentrations of several reducing materials, nutrients, and contaminants.

- Fourthly, many chemical reactions occur in the interstitial water during transport, as well as during dumping, and after sedimentation and so after the interstitial water is expelled into the overlying water several variables will change due to chemical reactions.

Programme

To come to grips with the problems likely to be created by dumping mud into a lake, the processes operating during dredging and subsequently, during and after dumping, have to be considered. We originally thought of carrying out a trial dumping in a lake, but in the face of political opposition it was decided to make our experiment at the head of a deep part of the port of Rotterdam. We selected Rotterdam's Eighth Petroleum Harbour (Fig. 1), where water velocity does not exceed 3 cm/s and the depth is about 20 m. In this harbour three pits were dredged, in which mud with different contamination levels was dumped. This material was obtained from Waalhaven, Botlekhaven and Maasmond harbours (Fig. 1). The programme adopted for the experiment provided successively for: (Fig. 3)

- Firstly, dredging the pits.

- Secondly, determining the contaminant and nutrient concentrations in the surface water and in the interstitial water of the mud on the bottom of the harbour; also determining the contaminant concentration of the sediment of the harbours and of the lake.

- Thirdly, dumping the mud in the pits. During dumping turbidity was measured and also the concentration of contaminants and nutrients in the surface water.

- Fourthly, determining the changes occurring in the dumped mud; the contaminant and nutrient contents of the pits were measured immediately after dumping and again after specific intervals.

After these experiments a number of calculations and simulations were conducted:

- Firstly, calculation of the turbidity during dumping and after

WATER QUALITY ASPECTS OF DUMPING DREDGED SILT 177

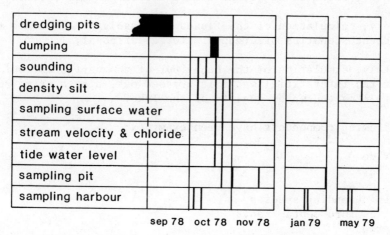

Fig. 3. Schedule for the measuring programme of Waalhaven mud.

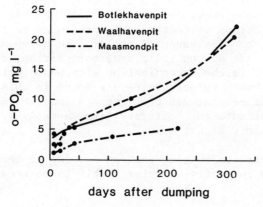

Fig. 4. Ortho-phosphate concentration in the interstitial water in the trial pits.

filling until a depth of 5 m of water had been attained;

- Secondly, calculation of the consolidation rate of the mud;

- Thirdly, calculation of the fluxes of dissolved components for different scenarios of dredged material disposal;

- Fourthly, simulation of the behaviour of the variables, which after the interstitial water is pressed out, play a prime role in controlling the water quality of the lake.

The following components have been examined:

- chloride
- nutrients
- heavy metals
- oxygen demanding species
- organic micro contaminants
- bacteria

EXPERIMENTAL RESULTS

Having outlined the programme of study we can now proceed to examine the results of the experiments.

Harbours and Lake

The study of the sediment and of the interstitial water in the harbours and in the lake gave the following results, (Table 1); Delft Hydraulic Laboratory Reports (1978, 1979). As can be seen the highest concentrations of contaminants are found in the mud of the eastern harbour, the Waalhaven (Fig. 1). The heavy metal content of sediments depends on the grain size composition, the smaller the grains the higher is the contamination with heavy metals. In the Rotterdam area linear relationships are found in this respect between metal content and percentages of fractions smaller than sixteen micrometers (Institute for Soil Fertility Report 1972; Delft Hydraulic Laboratory Report 1979).

It is however remarkable, that on looking at the concentrations dissolved in the interstitial water (Table 2) no great difference is apparent.

Dredging, Transport and Dumping

The density of the mud on the bottom of the harbour is too high

Table 1. Metal concentrations (in g/g.) in mud of the harbours and of the lake when 50% of the sediment by weight is < 16 µm.

	Maasmond	Botlekhaven	Waalhaven	Oostvoornse Meer
zinc	270	800	1500	300
copper	42	107	220	40
chromium	135	270	480	115
lead	80	180	320	85
cadmium	3.0	14	20	1.4
nickel	26	45	60	32
arseen	20	25	45	
mercury	0.80	2.2	5.5	

Table 2. Concentrations in interstitial water of mud in the harbours.

parameter	unit	Maasmond	Botlekhaven	Waalhaven
chloride	°/oo	13.0 - 15.3	5.5 - 6.0	1.6 - 2.3
sulphate	mg/l	500 - 2200	34 - 400	7 - 10
pH		7.2 - 7.3	7.0 - 7.2	7.0 - 7.1
ortho-phosphate	mgP/l	1.5 - 8.8	5.7 - 6.4	2.5 - 3.5
ammonia	mgN/l	19 - 56	49 - 70	57 - 90
silica	mg/l	5.0 - 8.9	5.7 - 6.9	9.4 - 10
iron	mg/l	17 - 29	24 - 38	20 - 37
manganese	mg/l	8.3 - 9.4	2.8 - 5.3	1.7 - 2.3
zinc	µg/l	11 - 30	11	11 - 14
copper	µg/l	4.0 - 6.4	2.0 - 4.0	2.0 - 4.2
lead	µg/l	2.4 - 13	2.4 - 11	2.4 - 5.4
nickel	µg/l	17 - 35	13 - 23	5.0 - 13
cadmium	µg/l	0.5 - 2.1	0.5 - 2.6	0.5 - 1.4
chromium	µg/l	1.7 - 6.0	4.1 - 9.8	4.5 - 9.3
arsenic	µg/l	34 - 58	6.3 - 11	2.4 - 4.9
fluoride	mg/l	1.8 - 2.6	1.9 - 2.0	0.5 - 0.6
Biochemical Oxyd. Demand	mgO$_2$/l	14 - 29	15 - 18	11 - 17
Dissolved Organic Carbon	mgC/l	33 - 81	37 - 49	34 - 41

for hydraulic transport, and so water from the harbour-basin is added. During dredging, alterations take place as a result of mixing of aerobic surface water with anaerobic mud and also after dumping many reactions which result in a change in the concentration of many chemical species occur (Berner 1974; Hartmann, et al., 1973; Schink, et al., 1975; Borght, et al., 1977.)

In order to investigate these alterations three kinds of mud were dumped in three trial pits. The size of the pits was 80 m x 30 m; the depth was 7 m. Concentrations of various species in the interstitial water in the silt were determined at different times, over a period of almost one year. During this period the rate of consolidation was also followed.

As an example of these alterations, Fig. 4 shows that the path of the concentration of ortho-phosphate, as a result of the decay of organic matter and reduction reactions, increases with time (Delft Hydraulic Laboratory Report 1980).

The sulphate concentration, on the contrary, remains almost constant during the first two months, and afterwards it decreases in all kinds of mud. Iron increases to a maximum concentration after which the concentration declines.

After dumping, a thick mud layer lies on the bottom of the lake. The density of the silt in the pit after dumping was 1040 to 1275 Kg/m^3. Consolidation takes place i.e. the original thickness of the layer, H_o, decreases with time by ΔH and the interstitial water is pressed out (Municipality of Rotterdam Report 1980). In Fig. 5 the rate of consolidation of the different kinds of mud is shown, the amount of water pressed out of the silt being clearly a function of time.

Experiments on a laboratory scale (Municipality of Rotterdam Report 1980) show a good similarity with the experiments in the pits. Density in the pits was measured with backscatter-sondes.

Thus the flux of dissolved components depends on the rate of consolidation as well as on the concentration in the pressed out pore-water. Both consolidation rate and concentration in the interstitial water are functions of time.

CALCULATIONS

From the results of these experiments three different things were modelled: turbidity, fluxes and reactions.

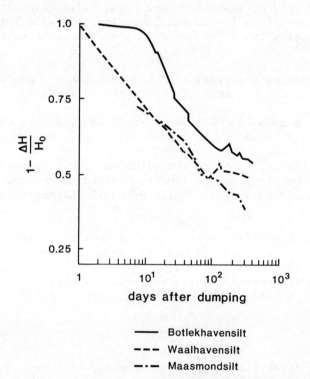

Fig. 5. Consolidation curves of different kinds of mud.

Turbidity

The turbidity of the lake early in 1978 was 17 mg/l (Delft Hydraulic Laboratory Report 1978). During and after dumping, the turbidity will increase. The level of turbidity depends mainly on water depth and wind velocity. The settling velocity of the sediment is an added factor and after sedimentation turbidity may increase as a result of increased erosion. Therefore the critical entrainment velocity, as well as the scour velocity of the settled sediment is important, (Migniot, 1968; Cormault, 1971 and Partheniades, 1962.) The settling velocity of silt depends on the type of sediment as shown by analyses of (Harrison and Owen, 1971; Rijkswaterstaat, Directie Waterhuishouding en Waterbeweging, District Zuidwest Report 1976). The salinity of the water is also important as is shown in Fig. 6.

When studying the erosion of the lake bottom, one should bear in mind that there are two main types of sediment:

- either the bottom is composed of rigid material, in which case the critical scour velocity is important;

- or the sediment is not yet consolidated, in which case there is hardly any critical scour velocity because the bottom is merely liquid mud (Sprong, 1971; Delft Hydraulic Laboratory Report 1974).

The curves in Fig. 7 illustrate what happens when material is poured into a lake while the wind is freshening and the depth of the water is 5 - 10 metres. Mean wind velocity is about 6.5 m/s. The turbidity under average conditions is 22 - 45 mg/l. This is the situation which holds after the sediment has consolidated.

Fluxes

Dumping takes place during 30 weeks of each year. During this period much interstitial water is pressed out and the concentrations of dissolved components are rather low. After this the rate of consolidation decreases and the concentration of the various pollutants increases. Table 3 illustrates this for silt which has been dumped into a pit in the Mississippi harbour and remained there for 9 weeks before being transported to the lake (Delft Hydraulic Laboratory Report 1980). This pit is called a "buffer".

In addition the influence of the components of the porewater flux on water quality in the lake is important. Some of the components change after they have been pressed out. One of the conservative components is chloride. The concentration of chloride in the lake was computed assuming that the porewater expelled mixes fully

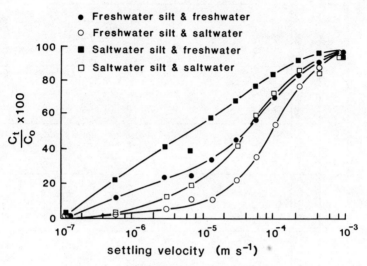

Fig. 6. Settling velocity of silt as a function of the salinity of the water and origin of the silt.

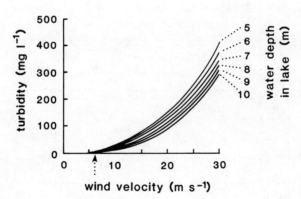

Fig. 7. Influence of wind velocity on turbidity in the lake (⇡ = mean wind velocity).

Table 3. Fluxes of dissolved components from the consolidating mud into the overlying lake water per week in the first, the thirtieth and fiftysecond week of the second year following dumping. Density during transport 1150 kg/m³.

Origin of mud parameter	unit	Waalhaven			Botlekhaven			Maasmond		
		week 1	week 30	week 52	week 1	week 30	week 52	week 1	week 30	week 52
o-PO$_4$	10^3 kg P	0.30	0.46	0.21	0.41	0.60	0.13	0.38	0.57	0.15
NH$_4$	10^3 kg N	1.7	2.5	0.49	0.55	0.81	0.32	1.7	2.5	0.84
Si	10^3 kg	0.34	0.52	0.25	0.25	0.36	0.16	0.59	0.89	0.40
Fe	10^3 kg	0.3	4.6	0.23	2.0	3.0	0.17	3.6	5.3	0.60
Mn	10^3 kg	0.14	0.21	0.022	0.15	0.22	0.018	1.1	1.7	0.16
As	kg	1.4	2.1	0.045	0.73	1.1	0.029	5.9	8.9	1.1
Cr	kg	0.22	0.33	0.19	0.14	0.20	0.12	0.24	0.36	0.081
Cu	kg	0.28	0.42	0.018	0.18	0.27	0.028	0.47	0.71	0.066
Pb	kg	0.48	0.73	0.13	0.37	0.54	0.055	1.3	2.0	0.24
Cd	kg	0.52	0.79	0.025	0.13	0.19	0.012	0.14	0.21	0.021
Ni	kg	1.7	2.5	0.14	1.6	2.3	0.19	4.7	7.1	0.24
BOD$_5$	10^3 kg O$_2$	0.96	1.5	0.23	0.37	0.54	0.18	1.7	2.5	0.69
DOC	10^3 kg C	2.1	3.1	0.59	1.4	2.0	0.39	6.6	9.9	1.1
oil	10^3 kg	0.021	0.032	0.011	0.027	0.040	0.001	0.019	0.029	0.003
phenol	kg	1.1	1.7	0.40	0.73	1.1	0.11	1.2	1.8	1.5

throughout the lake. Table 4 shows the result of the calculation for different origin muds with or without a 9 week stay in the "buffer" in the Mississippi harbour.

Reactions

In the previously mentioned calculations we have only considered the fluxes and the expulsion of conservative components. It will be clear that after the expulsion of the porewater, a lot of chemical reactions occur. We therefore decided to simulate the chemical alterations of several components in the lake water (Clasen, 1965; Shapely, et al., 1969).

With the aid of the computer programme Charon, (Rooij, 1980) we simulated the further chemical alterations of a number of components, which determine the chemical balance, the oxygen budget and the growth of algae.

Table 4. Calculated salinity of the lake following dumping of different muds. Water depth 5 m. (n.b. means 'no buffer' used; see text).

density during transport to Mississippi harbour (kg/m^3)	weeks in buffer-pit	density during transport to Oostvoornse Meer kg/m^3	Waalhaven mud chloride (°/oo)	Botlekhaven mud chloride (°/oo)	Maasmond mud chloride (°/oo)
1150	9	1150	12.0	12.4	13.1
1150	9	1200	11.7	12.1	13.2
1200	9	1150	12.4	12.6	13.2
1200	9	1200	12.1	12.4	13.3
1150	17	1150	12.3	12.6	13.2
1150	17	1200	11.8	12.2	13.2
1200	17	1150	12.6	12.8	13.3
1200	17	1200	12.3	12.5	13.4
n.b.	n.b.	1150	5.6	6.7	13.3
n.b.	n.b.	1200	6.6	7.6	13.5

The Charon model gives a numerical solution to the chemical equilibrium problem, by simulating the equilibrium of components. It distinguishes between the components with a high reaction rate that transform quickly, and the components with a low reaction rate. The first module calculates the distribution of components over combinations with a high reaction rate.

The most important processes for the second module are:

- transport of O_2, CO_2 and NH_3 between air and water (Delft Hydraulic Laboratory Report 1978)

- transport of porewater pressed out into the overlying water (Delft Hydraulic Laboratory Report 1980)

- oxidation of organic matter, (Delft Hydraulic Laboratory Report 1980)

- nitrification and denitrification

- precipitation of iron-hydroxide and iron-phosphate.

All the different components which might be important were included in the equilibrium calculation such as oxygen, ammonia, nitrate, phosphate, iron, biochemical oxygen demand and dissolved organic carbon.

The modelling of the several variables in the water involved recognition of the following:-

- the origin of the dredged material
- the density of the silt during transport
- the yearly amount of dumped silt.

Besides the variation of the above mentioned parameters the possible effects of thermal stratification during the first four years were modelled, after which the lake became too shallow for thermal stratification to occur. As the wind conditions and the temperature are very important for reaeration and oxygen conditions in the lake, in addition to the average values, the most unfavourable ones were all taken into account (very low wind velocities, low temperatures in winter and high temperatures in summer).

The results of the calculations, (Delft Hydraulic Laboratory Report 1981), show that the pattern of the concentrations in the lake are the same for all concentrations of variables. Fig. 8, for example, gives the oxygen content in relation to time. As dumping takes place during only 30 weeks a year, the oxygen concentration

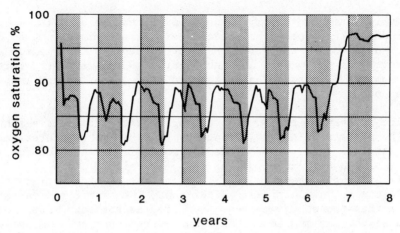

Fig. 8. Oxygen content in the lake water as percentage of the oxygen saturation during 8 year period, including 7 disposal years. Stippled areas show October to May dumping periods.

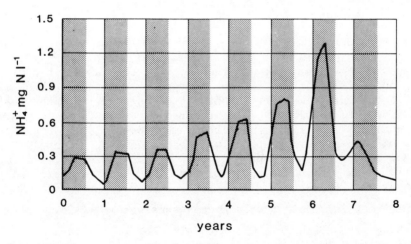

Fig. 9. Ammonia content in the lake water during 8 years including 7 disposal years. Stippled areas show October to May dumping periods.

is given as the percentage of the oxygen saturation, which is a function of temperature.

During dumping the oxygen content is reduced as a result of oxygen demanding biochemical reactions. In summer the amount of oxygen supplied by reaeration is greater than the consumption and so recovery of the oxygen content occurs. A return to the initial condition of 96% saturation is reached 30 weeks after the last dumping period.

The pattern of the NO_3^- content against time differs slightly from the pattern of the NH_4^+ content (Fig. 9). The interstitial water which is expelled from the bottom sediment into the surface water has high concentrations of NH_4^+ (ammonium). As a result of nitrification this becomes NO_3^-. During the dumping period, the ammonia concentration rises. After that, the nitrification rate is greater than the ammonium-flux. However, as the lake water volume decreases the maximum in the NH_4^+ concentration increases, since the fluxes from the sediment are equal every year. Neither NH_4^+ nor NO_3^- come to levels which are above the levels of the Dutch norm for water quality.

Due to the amounts of Fe^{2+} from the interstitial water, no phosphate problems occur during the dredged material disposal. However, in time, the Fe^{2+} concentration in the interstitial water decreases, whereas the PO_4^{3-} concentration increases. This implies that after the disposal has been completed, the PO_4^{3-} removing mechanism present in the system is low.

PRACTICAL MEASURES

As will be clear, the water quality of the lake is becoming worse, so it is important to look for possible measures which might be taken to improve matters. What can be done to limit the influence of dumping into the lake? What measures might be taken to guard the water quality? The first, as already mentioned, is to dump only during winter; all calculations were based on that principle. There are, however, a few more measures that could be devised. The most effective ones seem to be the following (MKO Report, Municipality of Rotterdam 1981):-

- dumping may influence turbidity and thus water quality
- flushing with clean salt water to give lower concentrations owing to dilution
- water depth might have some influence
- rate of infilling

Method of Discharging Spoil into the Lake

What is the best way to get the mud into the lake? Trial dumpings with clean mud in another lake, (Municipality of Rotterdam Report 1977), showed that turbidity is minimal when the spoil is dumped under the water-surface. It is important not to mix the mud with surrounding water, and so the end of the pipe must be near the bottom. But there is a great risk that the mud which is already becoming consolidated will be eroded and stirred up if too much energy is added (Abraham, 1963). Thus the end of the pipe must be constructed in such a manner that the dumped mud flows out with very low velocity.

The next aspect of the filling technique is the amount of diluting water used during dredging. There are two reasons to use as little diluting water as possible. First this water itself is not clean; secondly using less water means less interstitial water is pressed out during consolidation. By increasing the density of the mud discharged into the lake the amount of interstitial water per cubic metre of dredged spoil, which is squeezed out by consolidation, decreases substantially. In all cases the density after consolidation is 1300 kg/m^3.

Flushing

Part of the lake water may be pumped out and replaced by clean seawater. This proves to be the best way to improve the water quality (Fig. 10). Contaminants are not only removed in this way, but there is also a second positive effect, the addition of extra oxygen. Due to this the decay of organic material is promoted, and the risk of oxygen deficiency is diminished. Further there is increased chlorinity - which is important for the organisms which live in the lake and which are dependant on a high chlorinity in the water. As most of the dredged spoil comes from brackish or fresh water it is important to rinse through with salt water in order to keep chlorinity at the desired level.

Water Depth

The final depth of the lake must now be considered. Calculations had been carried out assuming a final depth of 5 m, but it might have been better to choose another depth. When depth is decreased the influence of turbulence on the bottom is more important. Mud from the bottom will then be stirred up and turbidity will increase. When depth is greater the influence of turbulence decreases, but calculations indicate that the effect of choosing a greater depth is negligible (MKO Report 1981).

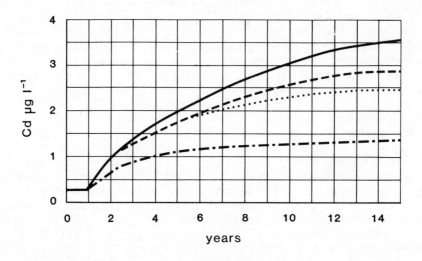

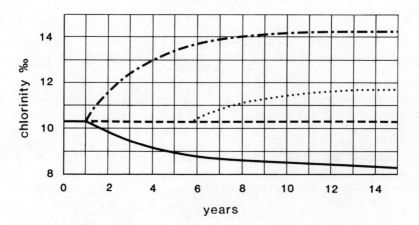

——— no flushing   ······ flushing − $10^5 \, m^3 \, week^{-1}$
– – – flushing − $5 \times 10^4 \, m^3 \, week^{-1}$ –·–·– flushing − $3 \times 10^5 \, m^3 \, week^{-1}$

Fig. 10. The influence of flushing on cadmium concentrations and chlorinity of the lake water.

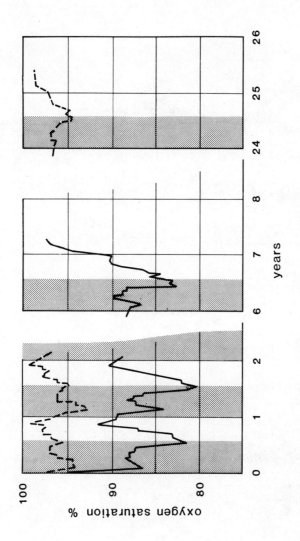

Fig. 11. Difference in oxygen levels in the lake water between 7 years disposal and 25 years disposal. Stippled areas show October to May dumping periods.

DREDGING

little dilution water
high density

TRANSPORT

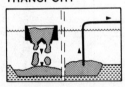

high density
temporary storage ("buffer")
little dilution water

DUMPING

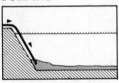

under water
low discharge velocity
high density
flushing
dumping period

AFTERWARDS

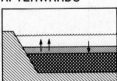

quality control
flushing
final water depth

Fig. 12. Aspects of dredging and disposal which need close control.

Also other considerations have to be taken into account in relation to benthos and diving waterbirds (Projectgroep Brielse Gat Report 1978). In view of these requirements a water depth of 5 m appears to be optimum.

Dumping Period

We have considered a number of other ways of diminishing the effect of the dumping. (MKO Report 1981). Most of them proved to be of little value, but one more should be mentioned: the period over which dumping takes place. When we dump for only a few years, say 7, the problems will be limited to that period, but in the meantime the water quality is poorer. Dumping over, say 25 years, maintains a better water quality, but we have to accept mediocre water quality for a longer time. To illustrate this, see Fig. 11 for the oxygen content.

CONCLUSIONS

Dumping of dredged material into a lake will reduce the quality of the water. Most of the variables reach the most unfavourable concentration during the last year of dumping. The quality of the lake water never becomes so bad that problems are to be expected in relation to the oxygen budget, growth of algae and pH.

An advantage of dumping into lakes is that one can continuously monitor and control the quality of the water. The measures to keep the water quality in hand are simple (Fig. 12). One of the most efficient measures appears to be flushing and this is neither expensive nor difficult.

REFERENCES

Abraham, G., 1963, Jet diffusion in stagnant ambient fluid, Delft Hydraulic Laboratory pub. 29.
Berner, R.A., 1974, Kinetic models for the early diagenesis of nitrogen, sulfur, phosphorus and silicon in anoxic marine sediments, in: The Sea, E.D. Goldberg, ed., 5:427.
Borght, J.P. van der, Wollast, R., and Billen, G., 1977, Kinetic models of diagenesis in disturbed sediments, Part 1 Mass transfer properties and silica diagenesis, Part 2 Nitrogen diagenesis, Limnol. Oceanogr. 22(5):787.
Clasen, R.J., 1965, The numerical solution to the chemical equilibrium problem, Rand Corporation RM-4345.
Cormault, P., 1971, Determination experimentale du debit solide d'erosion de sediments fins cohesifs, Proceedings 14th Congress of I.A.H.R., Paris.

Delft Hydraulic Laboratory, 1974, Menging bij een twee-lagensysteem veroorzaakt door wind Verslag modelonderzoek M 1235.
Delft Hydraulic Laboratory, 1976, Enkele beschouwingen over een proefstorting van baggerspecie in het Oostvoornse Meer, R 1141.
Delft Hydraulic Laboratory, 1978, Bemonsteringsopzet Oostvoornse Meer september 1978 - mei 1979.
Delft Hydraulic Laboratory, 1978, Natuurlijke beluchting van open water ten gevolge van wind, R 1318-II.
Delft Hydraulic Laboratory, 1979, Storten van baggerspecie uit de Botlek- en de Waalhaven in het Oostvoornse Meer, Interimverslag M 1501.
Delft Hydraulic Laboratory, 1979, Storten van baggerspecie uit de Europoort en de Maasmond in het Oostvoornse Meer, Interimverslag M 1549.
Delft Hydraulic Laboratory, 1979, Kwaliteitsonderzoek van baggerspecie in het Rotterdamse havengebied in 1977, M 1500.
Delft Hydraulic Laboratory, 1980, Microbial decomposition of organic matter and nutrient regeneration in natural waters and sediments. Report on literature study R 1310.
Delft Hydraulic Laboratory, 1980, (Groningen) Milieuchemische aspecten van het storten van baggerslib uit de Maasmond, de Botlek- end de Waalhaven in het Oostvoornse Meer M 1501.
Delft Hydraulic Laboratory, 1981, "Waterkwaliteitsaspecten tijdens en na het storten van baggerspecie uit de Maasmond, de Botlek- en de Waalhaven in het Oostvoornse Meer", M 1501.
Harrison, A.J.M., and Owen, M.W., 1971, Siltation of fine sediments in estuaries, Proceedings 14 Congress of I.A.H.R., Paris.
Hartman, M., Muller, P., Suees, E., and Weyden, C.H. van der, 1973, Oxidation of organic matter in recent marine sediments, Meteor. Forsch. Ergebnisse, Reihe C, no. 12:74.
Institute for Soil Fertility (Groningen), 1972, Onderzoek naar het voorkomen van zware metalen in bodemslib uit het Rotterdamse havengebeid.
MKO, 1981, Technische-, financiële- en waterkwaliteitsaspecten van verondieping van het Oostvoornse Meer met onderhoudsbaggerspecie R 80. Published by the Municipality of Rotterdam.
MKO, 1981, "Overzicht van de aspecten van verondiepen van het Oostvoornse Meer met baggerspecie", Published by the Municipality of Rotterdam.
Migniot, C., 1968, Etude des propriétés physique de différents sediments très fins et de leur comportement sous des actions hydrodynamiques, La Houille Blanche, 23, no. 7.
Municipality of Rotterdam, 1975, Het opvullen van het Oostvoornse Meer met havenslib 110.44-R75.35.
Municipality of Rotterdam, 1977, Proefstorting Zevenhuizerplas, 103.11 - R 7702.
Municipality of Rotterdam, 1980, Slibconsolidatie onder water, 59.00 - R 80.34.
Partheniades, E., 1962, A study of erosion and deposition of cohesive soils in salt water, Thesis Univ. of California.

Projectgroep Brielse Gat, 1978, Aspecten van verondieping van het Oostvoornse Meer met onderhoudsbaggerspecie.
Rijkswaterstaat, Directie Waterhuishouding en Waterbeweging District Zuidwest, 1976, Valsnelheid suspensief sediment, Meetresultaten monstername Rotterdamsche Waterweg - najaar 1975, nota nr. 44.003.01.
Rooij, N.M. de, 1980, A chemical model to describe nutrient dynamics in lakes. in: Hypertrophic Ecosystems, J. Barrica and L.R. Mùr ed., Jurk Publishers.
Schink, D.R., Guinasso, N.L., and Fanning, K.A., 1975, Processes affecting the concentration of silica at the sediment-water interface of the Atlantic Ocean, J. Geophys. Res. 80(21):3013.
Shapely, M., Cutler, L., De Hoven, J.C., and Shapiro, N., 1969, Specifications for a new jacobian package for the Rand chemical equilibrium problems, Rand Corporation RM-5426-PR.
Sprong, T.A., 1971, Onderzoek van menging bij dichtheidsverschillen, Delft, T.H. afdeling Weg- en Waterbouwkunde.
Swart, C.J., 1977, Het opvullen van het Oostvoornse Meer, Afstudeerverslag T.H. Delft.

AUTOTROPHIC IRON-OXIDISING BACTERIA FROM THE RIVER TAMAR

F.J. Cameron[1], E.I. Butler[2], M.V. Jones[1] and
C. Edwards[1]

[1] Department of Microbiology, University of Liverpool
Liverpool, UK

[2] Marine Biological Association, The Laboratory
Citadel Hill, Plymouth, UK

INTRODUCTION

Bacteria are involved in the transformations of many elements. Their role in the cycles of carbon and nitrogen has been intensively investigated. Their involvement in the biogeochemical cycles of many other elements, especially metals such as iron and manganese, is poorly understood. Considerable interest is now being shown in these transformations because of their economic and ecological importance. This report describes some studies into bacterial iron oxidation and forms part of a wider study of geochemical cycling of iron and manganese in the Tamar estuary.

Micro-organisms are implicated in the precipitation and solubilization of iron in a number of distinctly different ways and the chemical form of the element can be affected by a variety of biological mechanisms. A limited number of microbial species are directly responsible for transforming major deposits of iron and these are often referred to as the "iron bacteria".

A recent review of this group defined these organisms as (1) bacteria able to grow at the expense of energy obtained from the oxidation of reduced iron; (2) bacteria producing precipitates of iron generally associated with the cells; (3) bacteria causing reduction of ferric ions (Lundgren and Dean, 1979).

The first group includes the autotrophic iron-oxidizing bacteria. To date, few organisms have been reported to grow chemoautotrophically at the expense of Fe^{II}. Thiobacillus ferrooxidans is the only example which has been extensively studied. This Gram-negative, non-sporing, motile rod which was originally isolated from mine drainage water (Colmer et al., 1950), can oxidise ferrous sulphate, elemental sulphur, thiosulphate and many iron containing minerals (Silverman and Ehrlich, 1964). Generally, the following equations describe the oxidation of ferrous ions by T. ferrooxidans:

i. $\quad 4\ FeSO_4 + O_2 + 2H_2SO_4 = 2\ Fe_2(SO_4)_3 + 2H_2O$

ii. $\quad 2\ Fe_2(SO_4)_3 + 12H_2O = 4\ Fe(OH)_3 + 6H_2SO_4$

Gallionella ferruginea, a Gram-negative, microaerophilic bacterium, is widely distributed in iron-containing waters and often associated with ochre deposits. Maximum growth in artificial media, using ferrous sulphide as a source of reduced iron, is obtained at pH 6.3 to 6.6 (Kucera and Wolfe, 1957). In order to compete effectively with non-biological oxidation, this organism has to establish itself at a propitious point between the source of ferrous ions and air. This is manifested by a localized growth zone in laboratory enrichment cultures (Wolfe, 1964). Both Gallionella and Thiobacillus are capable of making considerable contributions to the transformation of iron in freshwater and terrestrial habitats.

An acid-tolerant, iron-oxidizing organism resembling Metallogenium spp was isolated by Walsh and Mitchell (1972). The organism repeatedly grew as multi-branching colonies or iron-encrusted filaments without a conventional cell body. However, recent investigations (Schmidt, cited in Reinheimer, 1980, Gregory et al., 1980) indicated that naturally-occurring material resembling "Metallogenium" may consist of non-living organic matter. The environmental significance of this organism must be questioned in the light of these findings. Table 1 summarizes some of the characteristics of these and other iron oxidizing bacteria isolated to date. This list does not include the thermophilic iron oxidizers.

In addition to the autotrophs, many heterotrophic bacteria may be involved in the transformations of iron (Cullimore and McCann, 1977). Enterobacter aerogenes and Serratia indica, amongst others, cause iron precipitation from organic salts by the utilization of the organic moiety. In the Sphaerotilus-Leptothrix group, the exact role of the iron oxidation process is not clear. It seems unlikely, however, that these organisms are able to derive energy from the reaction (see van Veen et al., 1978). It is known that in the case of Sphaerotilus natans, high concentrations of iron are growth-inhibitory (Chang et al., 1980). Many other organisms which have been reported to be associated with iron precipitates, e.g. Crenothrix, Lieskeella and Clonothrix, have not been grown in pure culture.

Table 1. Comparison of the basic characteristics of some iron-oxidising bacteria

	Shape	pH Range for growth	Mn (II) oxidation	Autotrophic
Gallionella ferruginea (Kucera & Wolfe 1957)	'Bean' shaped cell-twisted stalk	6 – 7.5	–	+
Thiobacillus ferrooxidans (Vishniac, 1974)	Rod 1.0 x 0.5 μm	1.4 – 6.0	–	+
Metallogenium spp (Walsh & Mitchell, 1972)	Filament 0.1 – 0.4 μm in diameter	3.5 – 5.0	+	+
Leptothrix spp (van Veen et al., 1978)	Sheath containing rodshaped cells	6.0 – 7.5	+	–
Leptospirillum ferrooxidans (Balashova et al., 1974)	Vibroid cells	2.0 – 3.0	–	+
Sheffield Isolate (Cameron et al., 1981)	Filament 1.5 – 3 μm diameter	2.0 – 4.5	–	+

Very little information is available on the types of bacteria actively involved in iron oxidation in estuarine environments. Basic data is required before any realistic assessment of their contribution to metal cycling can be made. A recent development is the realization that the chemical state of iron can affect the solubility of other elements in the environment. Co-precipitation of phosphate and iron bound to the surface of bacterial microcolonies may affect algal productivity in Lake Washington (Gregory, et al., 1980). This is also of importance when considering the cycling of toxic chemicals and environmental radioactivity. A greater understanding of the role of bacteria in iron cycling is required. Here, we describe some of the factors which affect autotrophic iron-oxidizing bacteria in an estuarine environment, some of the physiological characteristics of the process and their distribution during seasonal variations.

AUTOTROPHIC IRON-OXIDIZING BACTERIA FROM THE RIVER TAMAR

Isolation and Properties of Iron Bacteria

The estuary of the River Tamar drains a metalliferous area in south west England, where, until the turn of the century, metals including lead, tin, copper and silver were mined. The catchment area of the river and its tributaries, a total of 1.5×10^9 m^2, contains varying amounts of iron. A plan of the river is shown in Fig. 1 together with numbered sites which indicate the selected sampling stations. These were chosen to span a range of salinities from seawater (station 1) through decreasing salinity to freshwater (station 9). The freshwater/seawater interface, although dependent on the flow rate of the river and the height of the tide, generally occurs between stations 3 to 6.

Sediment samples and water samples taken from the river above and below Calstock were used to inoculate a mineral salts medium (Leathen et al., 1956) supplemented with filter-sterilized $FeSO_4.7H_2O$ (200 µg.ml^{-1}). The final pH of the medium was 2.5; at this value no chemical oxidation of iron occurs.

After three to four weeks the samples were assessed for the deposition of orange precipitates, which might be indicative of bacterial iron oxidation. Such samples were examined microscopically for bacteria and the orange deposit was repeatedly subcultured to obtain a pure culture. Purity was assessed microscopically, by the presence of a single colony type on iron-salts agar (Manning, 1975) and by the failure to obtain growth on a variety of complex media.

Iron oxidizing bacteria were isolated from seven sites. The characteristics of each isolate (numbered T1 to T7) were examined.

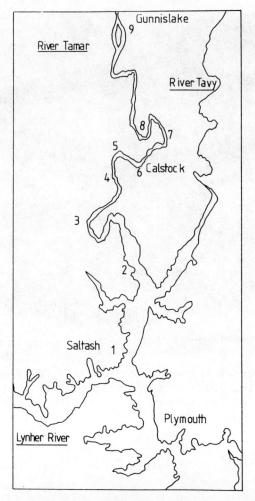

Fig. 1. The River Tamar. The numbered stations used as sampling sites extend from seawater at Saltash (Station 1), through to freshwater at Gunnislake (Station 9).

The optimum pH for iron oxidation was 2.0 for T2, T3 and T7, and 2.5 for the other isolates. Iron was oxidized from pH 1.5 to 4.5, above pH 4.5 chemical oxidation and precipitation of the resultant ferric salts made measurements of biological oxidation unreliable. Iron was oxidized in media which contained up to 500 µg ferrous ions.ml^{-1}. Only one isolate (T1) was inhibited by penicillin G (50 µg.ml^{-1}). Chloramphenicol resulted in variable growth inhibition

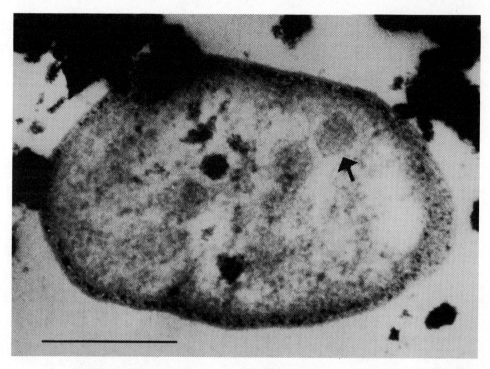

Fig. 2. Electron micrograph of a thin section of an isolated iron oxidizing bacterium. A peripheral ring of electron dense particles can be seen together with structures resembling carboxysomes (arrowed). Bar marker = 1 µ.

of the isolates. Nystatin (20 µg.ml^{-1}) and streptomycin (100 µg.ml^{-1}) were without effect.

All the isolates were rod-shaped organisms, approximately 1.5 µm by 0.8 µm. Transmission electron microscopy showed a peripheral layer of electron dense material (Fig. 2). Further examination using an energy dispersive analyser identified the major element present in this region of the cell as iron.

No other types of iron oxidizing bacteria were isolated using a variety of growth media and different cultural conditions.

Demonstration of the Autotrophic Nature of the Isolates

Areas in electron micrographs of the cells resembled carboxysomes. These are characteristic of autotrophic bacteria (Shiveley et al., 1973) and suggested that our isolates were also autotrophs. The isolates incorporated $[^{14}C]-CO_2$ when supplied as $NaH^{14}CO_3$ in the iron-salts medium. Initially iron oxidation occurred concurrently with incorporation of $[^{14}C]-CO_2$ into bacteria growing on glass fibre filters. Most chemoautotrophic bacteria utilize the ribulose 1, 5 diphosphate pathway for the assimilation of carbon (Whittenbury and Kelly, 1977). Using the assay of Lawlis et al., 1979, but with longer incubation times, it was shown that the incorporation of $[^{14}C]-CO_2$ could be greatly stimulated by ribulose 1, 5-diphosphate. This proved further evidence of the autotrophic nature of these isolates.

Coupling of Iron Oxidation to Respiration

The relationship between iron oxidation and oxygen consumption was measured for isolate T1 using a Gilson Respirometer. Appropriate controls showed that cell suspensions did not respire in the absence of Fe^{II}. In its presence, however, continuous oxygen consumption occurred until all the iron had been oxidized. Thereafter, respiration ceased. The amount of oxygen consumed for a given amount of iron could be readily calculated. Measurements of the amount of oxygen required to oxidize a range of amounts of iron indicated a stoichiometric relationship between the two processes (Fig. 3). From this data the proportion of iron oxidized to oxygen consumed could be calculated and expressed as the Fe^{II}/O ratio. We define this term as moles of Fe^{II} oxidized by moles of $1/2 O_2$. This ratio gave a value of 2.0 indicating that two electrons were released during a $2Fe^{II} \rightarrow 2Fe^{III}$ conversion, and probably reduced $1/2 O_2$ to H_2O via respiration. A proposed scheme for this process, together with a postulated site for ATP synthesis is shown in Fig. 4.

The reaction between $FeSO_4$ and O_2 has also been shown to be stoichiometric for the iron-oxidizing thiobacilli. In these bacteria rusticyanin functions as a periplasmic redox carrier to couple iron oxidation through cytochrome c to a cytochrome a_1 terminal oxidase (Silverman and Ehrlich, 1964; Dugan and Lundgren, 1965; Cox and Boxer, 1978). Energy coupling (to form ATP) occurs between cytochrome c and oxygen (Vernon et al., 1960).

Inhibition of Iron Oxidation by Sodium Chloride

It seemed pertinent to investigate whether NaCl would affect iron oxidation as the isolates came from habitats with fluctuating

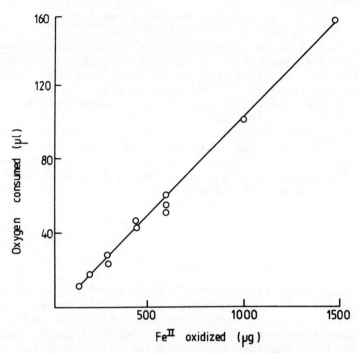

Fig. 3. The relationship between oxygen consumption and iron oxidized. Oxygen consumption was measured for fixed amounts of Fe^{II} using a Gilson respirometer. Appropriate controls showed no oxygen consumption in the absence of iron.

salinities. Iron oxidation by all the isolates was completely inhibited by NaCl at a concentration of 3% (w/v) which is approximately equivalent to that found in seawater. Lower concentrations gave lesser degrees of inhibition, and often caused a long delay before any oxidation was observed. Typical results for T1 are shown in Fig. 5. Adaptation to growth in the presence of intermediate concentrations of NaCl (1.5% w/v) was shown by a few of the isolates (Table 2). Oxygen uptake by all the isolates was also inhibited by the addition of NaCl to the iron-salts medium. With increasing NaCl concentrations there was a decrease in the Fe^{II}/O from 2.0 in the absence of NaCl to 0.6 in the presence of 2.5% (w/v) NaCl (Table 3). This suggested a decrease in the efficiency of the iron oxidation process caused by NaCl.

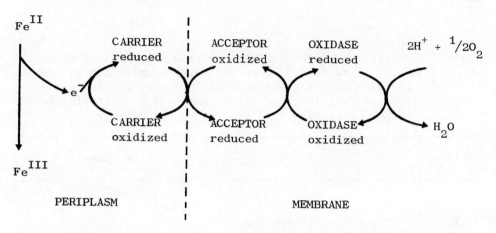

Fig. 4. Schematic coupling of iron oxidation to respiration. A site for ATP production is proposed between the redox carrier and the terminal oxidase of the respiratory chain.

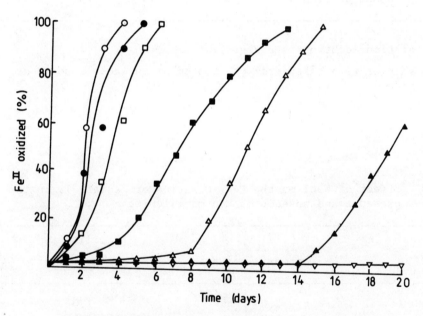

Fig. 5. Effect of NaCl on iron oxidation by isolate T1. The bacterium was inoculated into iron-salts medium containing different concentrations of NaCl: (○) 0% (w/v) NaCl; (●) 0.5% (w/v) NaCl; (□) 1% (w/v) NaCl; (■) 1.5% (w/v) NaCl; (△) 2% (w/v) NaCl; (▲) 2.5% (w/v) NaCl; (▽) 3% (w/v) NaCl. Samples were removed at daily intervals and the % oxidised iron determined.

Table 2. Adaptation of the isolates to growth in the presence of NaCl. Time (days) taken for 100% Fe oxidation by the isolates in the absence of NaCl (a), on initial exposure to 1.5% NaCl (b); and after continued subculturing in medium containing 1.5% NaCl (c).

Isolate	Source	Time (days)		
		a	b	c
T1	Ochre deposit station 7[a]	4	14	4
T2	Drain water station 7	4	21	21
T3	Mud[b] between station 6 & 7	4	16	12
T4	Mud between station 6 & 7	4	7	7
T5	Water station 6	4	12	8
T6	Mud between stations 5 & 6	4	12	4
T7	Mud station 5	4	20	–

[a] For station locations see Fig. 1.
[b] Mud samples were taken from the top cm of surfaces exposed at low tide.

Table 3. Effect of NaCl on the Fe^{II}/O ratios of isolate T1 in the presence and absence of Montmorillonite

	Fe^{II}/O − Montmorillonite	Fe^{II}/O + Montmorillonite
Control	2.0	1.0
+ NaCl 0.63%	1.8	1.1
1.25%	1.3	1.2
1.85%	0.7	1.3
2.50%	0.6	1.5

Interactions Between NaCl, Clay Minerals and Iron Oxidation

In view of the high content of particulate matter that can occur in this river (up to 1000 mg.l^{-1}) we examined the possible interactions between clay minerals and bacterial iron oxidation. Analysis of the suspended matter showed the presence of clay minerals (Table 4). Clay minerals have been shown to influence the growth of micro-organisms in a number of different ways (Stotszky, 1980). For example, survival of Escherichia coli in estuarine sediments was enhanced compared to the survival of the bacterium in seawater. A similar protection of the bacterium was observed in the presence of a montmorillonite clay (Roper and Marshall, 1979). In contrast iron oxidation by T. ferrooxidans was inhibited by a variety of particulate materials (Dispirito et al., 1981).

Examples of clays with high (montmorillonite) and low (kaolinite) cation exchange capacities and which are detectable in Tamar sediments (Table 4) were chosen for these experiments. A 1% (w/v) suspension of the clay in growth medium was used because higher concentrations adversely altered the pH. Lower concentrations necessitated the use of lower cell densities which made physiological measurements impracticable.

Montmorillonite caused a slight stimulation in the rate of iron oxidation whilst kaolinite resulted in slight inhibition. When the isolate was incubated in the presence of NaCl (1.5%, w/v) and montmorillonite (1%, w/v) the clay mineral afforded protection to the organism against NaCl toxicity. No oxidation occurred in the presence of NaCl (1.5%, w/v) plus or minus 1% (w.v) kaolinite (Fig. 6).

Table 4. Mineral composition of River Tamar suspended matter

Mineral	%
Kaolinite	11 - 24
Chlorite	17 - 23
Mica	25 - 51
Quartz	9 - 24
Montmorillonite	0 - 2
Tourmaline	0 - Trace

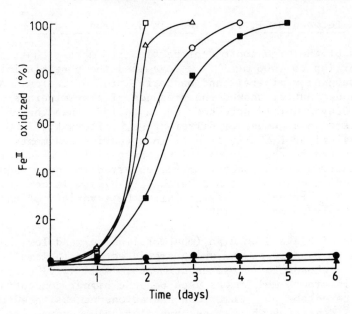

Fig. 6. Effects of NaCl and clay minerals on iron oxidation. Iron oxidation was determined with time in the presence and absence of NaCl, montmorillonite and kaolinite. No additions (control), (△); plus 1% (w/v) montmorillonite (□); plus 1% (w/v) kaolinite, (○); plus montmorillonite (1%, w/v) and 1.5% (w/v) NaCl, (■); plus kaolinite (1% w/v) and 1.5% (w/v) NaCl (●), plus 1.5% (w/v) NaCl alone (▲).

Montmorillonite has a higher cation exchange capacity than kaolinite and this is responsible for its protective effects (Stotszky, 1980).

Although the rate of iron oxidation in the presence of montmorillonite was not inhibited, the Fe^{II}/O was halved. This was due to an increase in oxygen consumption. In the presence of NaCl and montmorillonite the Fe^{II}/O increased from 1.1 at 0.63% (w/v) NaCl to 1.5 at 2.5% (w/v) NaCl (Table 3). This reflected less respiratory stimulation.

To investigate further the effects of montmorillonite on the metabolic activity of the isolates, the incorporation of $[^{14}C]-CO_2$ was determined. In the presence of montmorillonite there was a doubling of incorporation into cell material, although the amount of iron oxidized remained the same. We cannot explain this surprising result at present. However, it implies either a dissociation of oxygen consumption from iron oxidation, but with no loss of energy

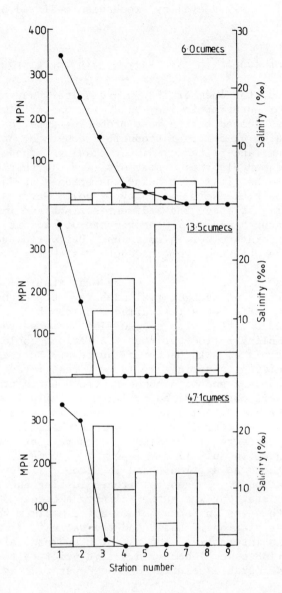

Fig. 7. Numbers of autotrophic iron-oxidizing bacteria recovered at the 9 stations during periods of (a) low (6/80) (b) medium (7/80) (c) high (10/80) flow rates. Mean daily flow rates for the 7 days prior to sampling are shown at the top right of each figure. (●) salinity at the time of sampling. Bacterial numbers per 100 ml were determined by a most probable number method (MPN).

generation, or an 'iron-cycling' mechanism mediated by the clay mineral.

DISTRIBUTION OF IRON-OXIDIZING BACTERIA IN THE RIVER TAMAR.

During June to November 1980 a preliminary survey of the distribution of autotrophic iron-oxidizing bacteria was undertaken. Initial studies showed that a most-probable number method using iron-salts medium gave more reproducible counts of bacterial numbers than either membrane-filter or direct plating methods. Water samples were collected monthly from 9 stations (Fig. 1) that covered the tidal range of the river from below the bridges at Saltash (Salinity 20º/oo) to Gunnislake Weir (Salinity 0º/oo). Water samples were collected from just above the sediment surface and immediately diluted into growth medium. Measurements of salinity, suspended matter and temperature were also taken. Daily flow rates at Gunnislake were obtained from the South West Water Authority.

Transects across the river at several stations did not show very great variations in bacterial numbers, but the numbers recovered from different sites varied considerably from month to month. Suspended matter also showed fluctuations from 20-1000 mg/l during the sampling period but these variations could not be positively correlated with bacterial numbers. During June and August 1980 iron-oxidizing bacteria were largely restricted to the upper reaches of the river and high numbers were counted at station 9. These months were characterised by low mean daily flow rates for the 7 day period preceding sampling (1.9 cumecs in June; 6.0 cumecs in August). When the river flow rates were higher, iron oxidising bacteria were recovered in greater numbers further down the river (Fig. 7). Hence, in July (7 day mean = 13.5 cumecs) the maximum numbers were found at station 3 and in October (7 day mean = 47.1 cumecs) at station 6. In general iron-oxidizing bacteria were not recovered from stations where the salinity exceeded 5º/oo. The only exception to this was in September when the maximum bacterial populations were at stations 2 and 3 and the salinity at station 2 was 8.6º/oo. It is interesting that the highest mean daily flow rate for the year (188 cumecs) occurred 3 days prior to the September sample and considerable scouring of the upper reaches was observed.

It is therefore proposed that the flow rate is a major factor affecting the distribution of iron-oxidizing bacteria in the River Tamar. Data obtained so far during 1981 supports this contention. It is also obvious from both laboratory and field studies that the salinity limits the populations of these organisms in the lower reaches either by directly inhibiting their growth or by precipitation of the bacteria associated with colloidal inorganic iron particles during estuarine mixing. It has been reported that where salinities are greater than 5º/oo iron is rapidly removed from solution (Morris et al., 1978). This factor would also affect the growth of iron bacteria.

The sources of iron and also iron oxidizing bacteria at any given station will be derived from that flowing in from above the site, that released from sediments and that washing in from the banks. In the region above Calstock (Station 5) the banks may be a major source of both iron and bacteria as drainage channels and seepage exhibit ochre deposits. A microscopical examination of these deposits reveals a mixed bacteria flora some of which may be involved in iron oxidation. A fuller characterisation of these microbial communities is being undertaken to try to establish whether they are a major source of the iron-oxidizing bacteria recovered from the river.

DISCUSSION

We have developed methods for the isolation of iron-oxidising bacteria from natural waters. All our isolates appear to be autotrophic bacteria capable of growth using CO_2 as the carbon source and energy derived solely from the oxidation of $Fe^{II} \rightarrow Fe^{III}$. From a thermodynamic viewpoint this would appear to be difficult to achieve. At pH 7 the standard redox potential (E'_o) of the Fe^{II}/Fe^{III} couple is +770 mV which is very close to that of the $H_2O/\frac{1}{2}O_2$ couple at +820 mV. The difference in standard free energy is approximately 2.3 Kcal.mol^{-1}. It has been calculated that the synthesis of ATP from ADP and inorganic phosphate requires an input of 7.4 Kcal.mol^{-1}, some three times greater than that allowed for from theoretical considerations of the Fe^{II}/Fe^{III} couple. It is well known, however, that membrane-bound redox carriers can exhibit widely differing redox potentials in vivo. Such membrane-associated phenomena may allow for a much greater standard free energy difference in the iron-oxidizing centres of our isolates. A possible involvement of membrane events is suggested by electron microscopy which shows iron particles within the outer layers of some of our isolates.

It is of considerable interest that NaCl inhibits iron oxidation. This implies that in an estuarine environment, bacterial iron oxidation will be most important in areas of low salinity. Such a proposal is supported by the fact that to our knowledge, no marine autotrophic iron-oxidizing bacteria have been isolated to date. Preliminary experiments in our laboratory suggest that it is a specific NaCl effect and not a Na^+ or Cl^- ion inhibition. Other chloride salts and sodium salts do not necessarily inhibit iron oxidation.

Iron oxidation by our isolates was maximal at low pH and therefore their activity in the estuarine hydrosphere will be limited. However, at sediment surfaces localized areas of very low pH have been observed (Butler, unpublished results). Bacterial metabolism in anoxic sediments will result in the release of organic acids and ferrous ions which can then be oxidized on the sediment surface. Bacteria can also produce ferrous ions by using ferric salts as an electron sink during anaerobic respiration (Ottow, 1968). Therefore significant bacterial iron oxidation will most likely occur at sediment surfaces in areas of low salinity.

The protection, and apparent enhanced iron oxidation which is observed in the presence of montmorillonite is difficult to explain with our present state of knowledge. In particular, the possibility of lattice substitution of aluminium by iron during bacterial iron oxidation may be an important factor and merits further investigation.

A strong correlation between water flow rate in the estuary and the distribution of iron bacteria was observed. Without further data it is not possible to decide whether this is due to (a) drainage into the river of more iron and/or bacteria or (b) the mobilization of sediments resulting in release of iron and/or bacteria. This will be the subject of further investigations.

ACKNOWLEDGEMENTS

We are grateful to the Natural Environment Research Council for financial support (GST/02/05,) and the help received from the crew of M.V. Gammarus with collection of samples.

REFERENCES

Balashova, V.V., Vedenina, L., Markosyan, G.E. and Zavarin, G.A., 1974, The auxotrophic growth of Leptospirillum ferrooxidans, Mikrobiol., 43:581.

Cameron, F.J., Edwards, C. and Jones, M.V., 1981, Isolation and primary characterization of an iron-oxidizing bacterium from an ochre polluted stream, J. Gen. Microbiol., 124:213.

Chang, Y., Pfeffer, J.T. and Chian, E.S.K., 1980, Distribution of iron in Sphaerotilus and the associated inhibition, Appl. Envir. Microbiol., 40:1049.

Colmer, A.R., Temple, K.L. and Hinkle, M.E., 1950, An iron-oxidizing bacterium from the acid drainage of some bitumous coal mines, J. Bacteriol., 59:317.

Cox, J.C. and Boxer, D.H., 1978, The purification and some properties of rusticyanin, a blue copper protein involved in iron (II) oxidation from Thiobacillus ferrooxidans, Biochem. J., 174:497.

Cullimore, D.R. and McCann, A.E., 1977, The identification, cultivation and control of iron bacteria in ground water, in "Aquatic Microbiology". F.A. Skinner and J.M. Shewan, ed., Academic Press, London.

Dispirito, A.A., Dugan, P.R. and Tuovinen, O.H., 1981, Inhibitory effects of particulate materials in growing cultures of Thiobacillus ferrooxidans, Biotechnol. Bioeng., Vol. XII:2761.

Dugan, P.R. and Lundgren, D.G., 1965, Energy supply for the chemoautotrophic Ferrobacillus ferrooxidans, J. Bacteriol., 89:825.

Gregory, E., Perry, R.S. and Staley, J.T., 1980, Characterisation, distribution and significance of Metallogenium in Lake Washington, Microbial Ecol., 6:125.

Kucera, S. and Wolfe, R.S., 1957, A selective enrichment method for Gallionella ferruginea, J. Bacteriol., 74:344.

Lawlis, V.B., Gordon, G.L.R. and McFadden, B.A., 1979, Ribulose 1,5-biphosphate carboxylase/oxygenase from Pseudomonas oxalaticus, J. Bacteriol., 139:287.

Leathen, W.W., Kinsel, N.A. and Braley, S.A., 1956, Ferrobacillus ferrooxidans : A chemosynthetic autotrophic bacterium, J. Bacteriol., 72:700.

Lundgren, D.G., and Dean, W., 1979, Biogeochemistry of Iron, in: "Biogeochemical cycling of mineral-forming elements", P.A. Trudinger and D.J. Swaine, ed., Elsevier, Amsterdam.

Manning, H., 1975, A new medium for isolating iron-oxidising and heterotrophic acidophiles from acid mine drainage, Appl. Envir. Microbiol., 30:1010.

Morris, A.W., Mantova, R.F.C., Bale, A.J. and Howland, R.J.M., 1978, Very low salinity regions of estuaries : important sites for chemical and biological reactions, Nature 274:678.

Ottow, J.C.A., 1968, Evaluation of iron-reducing in Aerobacter aerogenes, 2, Allg. Mikrobiol., 8:441.

Rheinheimer, G., 1980, "Aquatic Microbiology" 2nd edn., Wiley, Chichester.

Roper, M.M. and Marshall, K.C., 1979, Effects of salinity on sedimentation and of particulates on survival of bacteria in estuarine habitats, Geomicrobiol. J., 1:103.

Shiveley, J.M., Ball, F.L. and Kline, B.W., 1973, Electron microscopy of the carboxysomes (polyhedral bodies) of Thiobacillus neapolitanus, J. Bacteriol.,116:1405.

Silverman, M.P. and Ehrlich, H.L., 1964, Microbial formation and degradation of minerals, Adv. Appl. Microbiol., 6:153.

Stotzky, G., 1980, Surface interactions between clay minerals and microbes, viruses, and soluble organics, and the probable importance of these interactions to the ecology of microbes in soil, in "Microbial Adhesion to Surfaces", R.C.W. Berkeley, J.M. Lynch, J. Melling, P.R. Rutter, and B. Vincent, ed., Ellis Horwood, Chichester.

van Veen, W.L., Mulder, E.E. and Deinema, M.H., 1978, The Sphaerotilus-Leptothrix group of bacteria, Microbiol. Rev., 42:329.

Vernon, L.P., Mangum, J.H., Beck, J.V. and Shafia, F.M., 1960, Studies on a ferrous-ion oxidising bacterium II. Cytochrome composition, Arch. Biochem. Biophys. 88:227.

Vishniac, W.V., 1974, Thiobacillus Beijerinck 1904, in "Bergey's Manual of Determinative Bacteriology" 8th edn., R.E. Buchanan and N.E. Gibbons, ed., Williams and Wilkins, Baltimore.

Walsh, F. and Mitchell, R., 1972, An acid tolerant, iron oxidising Metallogenium, J. Gen. Microbiol., 72:369.

Whittenbury, R. and Kelly, D.P., 1977, Autotrophy : a conceptual phoenix, in "Microbial Energetics" Symp. Soc. Gen. Microbiol. 27, B.A. Haddock and W.A. Hamilton ed., Cambridge U.P., Cambridge.

Wolfe, R.S. 1964, Iron and Manganese bacteria in "Principles and applications in aquatic microbiology", H. Heukelekion and N.C. Dondero, ed., Wiley, New York.

NITROGEN CYCLING BACTERIA AND DISSOLVED INORGANIC NITROGEN

IN INTERTIDAL ESTUARINE SEDIMENTS

N.J.P. Owens* and W.D.P. Stewart

Department of Biological Sciences
The University, Dundee, UK

*Present Address:
Institute for Marine Environmental Research
Prospect Place, The Hoe, Plymouth, UK

INTRODUCTION

It is well documented that aquatic sediments are sites of intensive microbial activity and that the majority of the processes which occur there involve the breakdown of organic matter and the subsequent transformations of the products of decomposition. Estuarine sediments typically have higher interstitial water concentrations of dissolved inorganic nitrogen (DIN) than in the overlying water column, suggesting that sediments are sites of active nitrogen cycling.

Biological nitrogen cycling is predominantly carried out by physiologically distinct groups of micro-organisms in a series of well defined transformations. There are two broad, fundamental processes. Firstly, the mineralisation of organic to inorganic nitrogen and secondly, interconversions between the various redox states of inorganic nitrogen resulting from the decomposition processes.

In a previous paper (Owens et al., 1979) it was shown that in the Eden Estuary, Scotland, there was a close coupling between sediment nitrogen cycling processes and primary production. It is the aim of this paper to describe, more fully, the role of the sediment nitrogen cycling bacteria and their interactions with nitrogen speciation and concentrations in interstitial water.

MATERIALS AND METHODS

The Study Area

The estuary under investigation was the Eden Estuary, Fife, Scotland (Nat. Grid Ref. NO 475195), situated approximately 10 km south of the River Tay and approximately 6 km in length from the upper limit of tidal influence to the mouth in St. Andrew's Bay. The estuary drains completely at low tide exposing extensive mud and sand flats. At approximately MHWS level there are salt marsh fringes. The study was carried out on a transect chosen to incorporate the major features of the estuary, along which eight sampling stations were established. See Owens et al., (1979) for a more detailed account of the study site.

Chemical Analyses

Sediment samples for chemical analyses were obtained by hand at low tide. Interstitial and exchangeable inorganic nitrogen species were determined spectrophotometrically after their extraction by the methods of Bremner (1965). NO_2-N was determined by the method of Bendschneider and Robinson (1952) as was NO_3-N, after reduction to NO_2-N by spongy-cadmium (Mackereth et al., 1978). NH_4-N was determined according to Solorzano (1969).

Enumeration of Bacteria

This was carried out using the MPN technique by decade dilutions and five replicates. The medium of Meiklejohn (1965) was used for ammonifying bacteria, those of Alexander and Clark (1965) for NH_4^+ and NO_2^- oxidisers. The CPS medium of Collins (1963) supplemented with 2 gl^{-1} KNO_3 was used for total and denitrifying bacteria. The salinity of all media was 35 ^{o}/oo. A similar relative efficiency of recovery of the particular bacteria was assumed for each medium.

^{15}N Analysis

^{15}N was determined by mass spectrometry after release of N_2 from NH_4^+ in an evacuated Rittenberg tube (see Stewart, 1967).

Inorganic Nitrogen Concentrations and Speciation

Data were obtained over a thirteen-month period between March, 1977 and March, 1978 on the DIN concentrations in the interstitial water in the top 3 cm sediment at eight sites in the Eden Estuary. The mean concentrations of NO_3-N and NH_4-N for the period are presented in Table 1, as are DIN data for the water column.

TABLE 1. The mean annual concentrations and range of NO_3-N and NH_4-N in the interstitial waters at various sampling sites along a sediment transect and in the overlying water at high tide

SITE		NO_3-N $\bar{x}$	NO_3-N Range	NH_4-N $\bar{x}$	NH_4-N Range
1	Mature Salt Marsh	790.7	194.1 – 1003.6	361.1	39.1 – 650.7
2		753.2	101.4 – 953.2	448.3	200.4 – 749.4
3		412.2	105.7 – 1010.3	257.9	86.3 – 466.8
4	Pioneer Marsh	422.9	47.6 – 723.6	195.9	22.6 – 303.5
5		365.2	168.8 – 770.2	173.4	25.9 – 411.2
6		451.0	195.1 – 998.5	184.5	54.0 – 510.3
7	Mud Flats	450.2	91.0 – 753.4	216.9	40.3 – 581.8
8		498.5	100.2 – 886.4	230.3	64.5 – 400.9
FLOOD WATER		146.0	32.8 – 249.7	82.9	29.5 – 176.2

Units: $\mu g\ N\ l^{-1}$

Data are means of values obtained during the period March, 1977 – March, 1978. The values are based on samples taken to a sediment depth of 3 cm.

Water column samples were collected during the flooding tide after collection of the sediment samples. At all sites, NO_3-N and NH_4-N concentrations were greater in interstitial water than in the water column. NO_3-N varied between 365.2 $\mu g Nl^{-1}$ at a mudflat site (Site 5) to 790.7 $\mu g Nl^{-1}$ interstitial water at a mature salt marsh site (Site 1). The mean concentrations of NH_4-N were less than NO_3-N and varied between 173.4 $\mu g Nl^{-1}$ (Site 5) and 448.3 $\mu g Nl^{-1}$ at a second salt marsh site (Site 2).

The seasonal change in speciation of DIN is shown in Figure 1. It can be seen for sediment Sites 1 and 7 that there was a change from a DIN pool comprised predominantly of NO_3-N in April to one comprised mainly of NH_4-N during June with a return to the NO_3-N dominated condition during October and January. The tidal floodwater DIN pool measured over the same period showed a similar

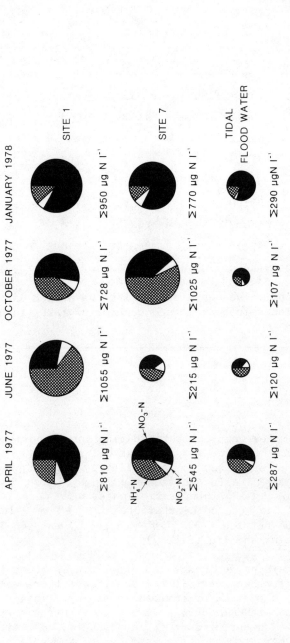

Figure 1. Relative abundance of NO_3-N, NH_4-N and NO_2-N in the interstitial water at two sediment sites and in tidal flood water.

pattern of speciation but NH_4-N comprised a maximum of only 50% of the total in comparison with the sediment sites where NH_4-N accounted for approximately 65% and 60%.

Seasonal Variation in Nitrogen Cycling Bacteria

Data obtained on the abundance of various nitrogen cycling bacteria at a mature salt marsh site (Site 1) and a mud-flat site (Site 8) are presented in Figure 2. Generally, for all the bacterial types enumerated, greater populations were found on the salt marsh than on the mud-flats.

At Site 1, aerobic heterotrophic bacteria (Figure 2a) showed little seasonal variation in abundance with the populations fluctuating between approximately 10^6 and 10^8 bacteria g^{-1} dry wt. sediment. The population of heterotrophs at Site 8 varied between 10^4 and 10^7 bacteria g^{-1} dry wt. sediment. There was some evidence of a seasonal variation at this site, with the population increasing from a minimum in March, to a maximum in September, followed by a decline during the autumn and winter.

Ammonifying bacteria, that is those capable of mineralising peptone nitrogen to NH_4^+, were the most abundant nitrogen cycling bacteria (Figure 2b) and on occasions comprised over 90% of the total heterotrophic bacterial population. Maximum numbers of ammonifiers occurred at Site 1 during winter and at Site 8 during autumn. The maximum at Site 8 coincided with the seasonal decline of Enteromorpha which had a high biomass on the mud-flats (see Owens et al., 1979).

Autotrophic NH_4^+ and NO_2^- oxidising bacteria (nitrifiers; Figure 2c and 2d) were present at both sites with populations of up to 10^4 bacteria g^{-1} dry wt. sediment but typically between 10^1 and 10^3 bacteria g^{-1} dry wt. sediment. Generally, highest numbers were found during autumn and winter although at Site 1 a large population of NH_4^+ oxidising bacteria was found during the summer.

Denitrifying bacteria (Figure 2e) showed the most marked seasonal variation in abundance, ranging from undetectable in August and September up to 10^3 bacteria g^{-1} dry wt. sediment during winter. The criterion used to distinguish denitrifying bacteria was the ability to produce gaseous by-products from NO_3^- utilization; the numbers found, therefore, were considerably less than in some studies where NO_3^- utilization was the criterion (see, for example, Jones, 1979).

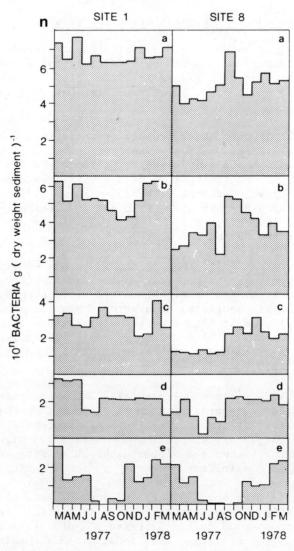

Figure 2. Seasonal variation in nitrogen cycling bacteria at two sediment sites in the Eden Estuary: (a) aerobic heterotrophic bacteria; (b) ammonifying bacteria; (c) NH_4^+ oxidising bacteria; (d) NO_2^- oxidising bacteria; (e) denitrifying bacteria.

NITROGEN CYCLING BACTERIA AND DISSOLVED NITROGEN

Rates of Nitrogen Cycling Processes

(a) <u>Ammonification</u>. Table 2 shows the results of experiments carried out to determine NH_4^+ release from the sediments. The data show that NH_4^+ was released in varying quantities throughout the year and maximised during autumn and early winter. Maximum NH_4^+ release coincided with the death and obvious decay of <u>Enteromorpha</u> and observations made on sediment cores at this time revealed considerable quantities of <u>Enteromorpha</u> buried within the sediment. The Eh of the sediments at this time was low (between −100 mV and −300 mV), indicative of rapid microbial degradation.

(b) <u>Nitrification</u>. Data on nitrification rates and autotrophic nitrifier populations are presented in Table 3. Nitrifiers were more abundant in the salt marsh sediments than on the mudflats (see, also, Figure 2). The rates of nitrification were correspondingly higher in the salt marsh sediments, there being over an order of magnitude more nitrogen oxidised than on the mudflats. For further information on nitrification in the Eden Estuary, including the relative importance of autotrophic and heterotrophic nitrification see Christofi et al. (1979).

TABLE 2. Ammonium released from sediment cores collected from Site 7 into tidal flood water

DATE	RATE OF NH_4 − N RELEASE ($\mu g\ NH_4-N\ m^{-2}\ h^{-1}$)
15.4.78	81.2
21.6.78	351.8
13.9.78	649.6
20.11.78	514.2
14.2.79	54.1

Data were obtained by incubating sediment cores 6 cm diam. x approximately 30 cm length, for 24 hours under simulated <u>in situ</u> conditions. Cores were collected by hand immediately prior to tidal innundation in perspex tubes and immediately capped at both ends. Flood water was collected at the site and on return to the laboratory 100 ml was carefully decanted into the perspex tubes overlying the sediment surface. Water samples (5 ml) were removed for analysis after 30 mins and 24 hrs. Values are the mean of three replicate cores.

TABLE 3. Nitrification rate and number of autotrophic nitrifiers at four sites in the Eden Estuary

SITE	$NO_2^- + NO_3^-$ PRODUCED ($\mu g\ N\ g^{-1}$ dry wt. sediment h^{-1})	AUTOTROPHIC NITRIFIERS NH_4^+ OXIDISERS (No. g^{-1} dry wt. sediment)	NO_2^- OXIDISERS
1 (Salt Marsh)	0.109	1,300	1,200
2	0.101	8,300	2,600
7 (Mud-Flats)	0.018	150	15
8	0.020	22	960

Nitrification rate obtained by incubating sediment samples with an equal volume of sterile flood water with and without an inhibitor of autotrophic nitrification (N-serve, 1.0 mgl^{-1} final concentration) for 48h at $18^{\circ}C$ in the dark. 5 ml water samples were removed before and after incubation and analysed for NO_2^- and NO_3^-. Autotrophic nitrifiers were determined on aliquots of sediment removed before incubation.

TABLE 4. NO_3^- reduction at two sites in the Eden Estuary

SITE	END PRODUCT (nMoles $N_2\ 10\ cm^{-3}\ 24h^{-1}$)	
	N_2O/N_2	NH_4^+
1 (Salt Marsh)	648(129)	497(98)
7 (Mud Flats)	456(161)	763(104)

Sediments were collected in June for gaseous by-products experiments and August for NH_4^+ by-products experiment. In each case 7 μMole NO_3-N were added to 10 cm^3 cores. For N_2O/N_2, $^{14}NO_3$ was used and 0.1 atmosphere C_2H_2 added to an airtight incubation vessel. The products of denitrification were measured by gas-chromatograph. For NH_4^+ by-products incubation, $^{15}NO_3$(99.6 at %) was added. NH_4^+ formed was collected by distillation and analysed for ^{15}N content as described in Materials and Methods. Values are the mean of 5 determinations and values in parenthesis are standard deviations.

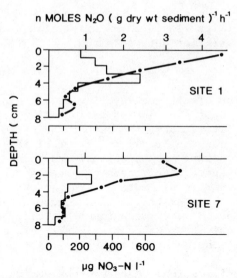

Figure 3. Depth distribution of NO_3-N concentrations (histograms) and N_2O production (._____.) in sediment cores from two sites in the Eden Estuary. N_2O production was estimated on cores with added acetylene (acetylene inhibition technique). Data were obtained on duplicate cores by excising the required segments and used either for extraction of nutrients or determination of N_2O by gas chromatography.

(c) Nitrate Reduction. The relationship between NO_3^- concentrations and denitrification is shown in Figure 3. The depth profiles show a NO_3^- maximum between 3-4 cm below which NO_3^- concentrations decreased. A gradual reduction in NO_3^- concentrations above 3-4 cm was also apparent and coincided with maximum denitrification rates.

The relative activities of NO_3^- dissimilation to gaseous by-products (denitrification) and NH_4^+ are compared in Table 4. Of the 7 μMole NO_3^- added to sediment cores, similar amounts were dissimilated by the two alternative metabolic pathways. With up to 9.25% denitrified (Site 1) and 10.9% dissimilated to NH_4^+ (Site 7). There was little variation between sites.

DISCUSSION

The enumeration of nitrogen cycling bacteria per se gives little information on the extent of nitrogen cycling. However, coupled with rate estimations, the two approaches may together provide information on the likely in situ activity of the process. In this study, there were marked spatial and temporal variations in the populations of specific nitrogen cycling bacteria and, combined with the evidence of rate estimations and the change in speciation and concentrations of interstitial DIN, suggests that the intertidal sediments of the Eden Estuary were important sites of bacterial nitrogen cycling activity. Collectively the data suggest an ordered sequence of the major nitrogen cycling transformations.

Firstly, ammonification. Enteromorpha was the major primary producer on the Eden mud-flats (see Owens et al., 1979) and represented a potentially large pool of organic carbon and nitrogen. Owens et al (1979) provided evidence, using ^{15}N labelled algae, of mineralisation of Enteromorpha in simulated in situ incubations. The seasonal variation in NH_4^+ release from the mud-flats (Table 2), which maximised at the time of the annual decline of the alga, suggests that in situ mineralisation of Enteromorpha may have been responsible, in part at least, for the observed sediment NH_4^+ concentrations and release rates. Maximum NH_4^+ release coincided with the highest populations of ammonifiers which formed a high proportion of the total heterotrophic bacterial population. The data are in general agreement with Jones (1971) and Little et al. (1979) who demonstrated maximum proteolytic bacterial activity following freshwater phytoplankton blooms and is consistent with many estuarine and marine sediment studies which have shown NH_4^+ to be the dominant pore water DIN species (see, for example, Blackburn, 1979; McCaffrey et al., 1980). The maximum number of ammonifiers occurred during winter on the salt marsh and possibly reflected the more refractory nature of the salt marsh angiosperms to bacterial mineralisation than Enteromorpha. An increase in NH_4^+ concentrations was, however, apparent in the salt marsh sediments and suggests that in situ mineralisation of organic nitrogen was occurring. The absence of NH_4 production or release data for this site does, however, emphasise the inadequacy of isolated bacterial population estimates.

In view of the high pore water NO_3^- concentrations, particularly during winter, and the increase in $NO_3^-:NH_4^+$ ratios, nitrification was indicated as the second 'phase' of nitrogen cycling in the Eden sediments. Nitrification is the only major biological nitrogen cycling process which increases the oxidation

state of nitrogen and is predominantly mediated through the
activity of the autotrophic NH_4^+ and NO_2^- oxidising bacteria.
The total numbers of these organisms recorded in the sediments
were as high or higher than those found in various agricultural
ecosystems (see, for example, Ashworth, 1973; Rowe et al., 1977).
Nitrifiers generally were more abundant during the winter months,
particularly on the mud-flats, and may possibly have been stimulated
by the summer and autumn increase in NH_4^+ concentrations.
Nitrifying bacteria in lakes, for example, have been shown to
respond to increases in NH_4^+ concentrations and unequivocally,
using ^{15}N, to nitrify previously ammonified nitrogen (Christofi
et al., 1981). Generally, the nitrification rates co-varied with
autotrophic nitrifier population size (Table 3) suggesting that
population estimates of nitrifiers may be used as a measure of, at
least, the nitrification potential (see also Christofi et al., 1981).
The rates of nitrification obtained were similar to those found by
Billen (1976) for sediments in the North Sea and by Koike and
Hattori (1978) for estuarine sediments in Japan and, as in these
studies, suggest that nitrification may be an important nitrogen
cycling pathway in estuarine sediments (see also Billen, 1975;
Vanderborght and Billen, 1975; Iizumi et al., 1980). There are
several important ecological consequences of nitrification.
Firstly, NO_3^- is more easily leached from sediments than NH_4^+,
which becomes bound onto sediments as exchangeable NH_4^+ (Rosenfeld,
1979); nitrification may therefore effectively reduce the nitrogen
status of the sediments by providing a route for tidal exchange
and loss. Secondly, NH_4^+ is generally the preferred DIN species
assimilated by algae (see McCarthy, 1980) and the reduction in the
$NH_4^+ : NO_3^-$ ratio as a result of nitrification may impose a
selection pressure and thus control the species composition of, for
example, the epipelic algal flora. Thirdly, NO_3^- in environments
of reduced oxygen tensions, conditions typical of estuarine
sediments, may act as an efficient terminal electron acceptor for
microbial oxidation of organic matter and in certain NO_3^- respiration pathways, NO_3^- may be dissimilated to gaseous by-products
(N_2O and N_2) thereby acting as a nitrogen sink (see Brezonik, 1977;
Painter, 1977). Data on the populations of denitrifying bacteria
indicate that the latter pathway may be important in the Eden
Estuary and may be the third 'phase' in the sediment nitrogen
cycle.

The denitrifying bacteria were present with populations
similar to those found by Todd and Nuner (1973) for agricultural
soils and Johnston et al. (1974) for freshwater sediments. The
seasonal variation was the most marked of all the types enumerated
and the maximum during winter, following the accumulation of NO_3^-,
suggests that denitrification of previously mineralized and
nitrified nitrogen could occur. As with the other bacterial types,
the presence of denitrifiers does not indicate _in situ_ denitrification (see, for example, Jones 1979; Jones and Simon, 1981).

However, Jones et al. (1980) demonstrated positive relationships between nitrate reductase activity and the denitrifying population in freshwaters and Chan and Campbell (1980) found denitrification rates in a lake in Canada were positively correlated with the number of denitrifying bacteria. The depth profiles shown in Figure 3 demonstrate the possible interaction between NO_3^- concentrations and denitrification. At both sites the reduction of NO_3^- concentrations at the sediment surface coincided with maximum denitrification rates and suggest that denitrification in these cores accounted for the observed NO_3^- profiles. Although denitrification proceeds most rapidly under reducing conditions, there is also evidence to suggest that the process is rate limited by NO_3^- supply and that maximum denitrification occurs in the more oxidised surface sediments close to the zone of maximum nitrification (see, for example, Vanderborght and Billen, 1975; Sorensen, 1978b, c). This is an apparent paradox as NO_3^- is only utilized as terminal electron acceptor when oxygen tensions are low. The interaction of oxygen concentrations and denitrification is, however, unclear but these data suggest that denitrification does occur at the sediment surface and possibly microsites of reduced oxygen tensions are involved (see, also, Brezonik, 1977; Focht and Verstraete, 1977; Jones, 1979; Jones et al., 1980).

The considerable reduction during the late summer in the number of denitrifying bacteria suggests that NO_3^- respiration is of little importance at this time of year. There is, however, recent evidence to suggest that an alternative pathway of NO_3^- dissimilation, with NH_4^+ as the end-product, may be important in aquatic sediments (see, for example, Dunn et al., 1978, 1979; Jones, 1979; Herbert, 1980; Jones and Simon, 1981). Using ^{15}N, we have confirmed the occurrence of NO_3^- reduction to NH_4^+ (Table 4) and, in agreement with Sorensen (1978a), found similar rates of both NO_3^- respiration pathways. The relative importance of denitrification and NO_3^- reduction is of considerable ecological importance as the former transformation acts as a nitrogen sink whilst the latter maintains the nitrogen status and as such warrants further study.

Overall, the data obtained in this study point to the sediments of the Eden Estuary as important sites of nitrogen cycling activity. Although there is evidence to suggest an ordered sequence of ammonification, nitrification and denitrification, there is no evidence that the nitrogen cycling transformations are mutually exclusive; but rather, the individual processes assume more importance during certain seasons and, together with interactions with the sediment DIN pools and tidal exchange, combine to enhance the nutrient status of the sediment ecosystem.

ACKNOWLEDGEMENTS

This work was carried out whilst N.J.P.O. was in receipt of a NERC Studentship.

REFERENCES

Alexander, M. and Clark, F.E. (1965). In: Methods of Soil Analysis (Part 2). (Ed. C.A. Black). American Society of Agronomy, Madison, Wisconsin. pp 1477.

Ashworth, J. (1973). On measuring nitrification and recovery of aqueous ammonia applied to grassland. J. Agric. Sci., Camb., 81; 145.

Bendschneider, K. and Robinson, R.J. (1952). A new spectrophotometric determination of nitrite in seawater. J. Mar. Res., 2: 87.

Billen, G. (1975). Nitrification in the Scheldt Estuary (Belgium and the Netherlands). Estuarine and Coastal Mar. Sci., 3; 79.

Billen, G. (1976). A method for evaluating nitrifying activity in sediments by dark (^{14}C)-bicarbonate incorporation. Water Res., 10: 51.

Blackburn, T.H. (1979). Method for measuring rates of NH_4^+ turnover in anoxic marine sediments, using a $^{15}N-NH_4^+$ dilution technique. Appl. Environ. Microbiol. 37; 760.

Bremner, J.M. (1965). In: Methods of Soil Analysis (Part 2) (Ed. C.A. Black). American Society of Agronomy, Madison, Wisconsin, pp. 1179.

Brezonik, P.L. (1977). Denitrification in natural waters. Prog. Wat. Technol., 8: 373.

Chan, Y.K. and Campbell, N.E.R. (1980). Denitrification in Lake 227 during summer stratification. Can. J. Fish. Aquat. Sci., 37; 506.

Christofi, N., Owens, N.J.P. and Stewart, W.D.P. (1979). Studies of nitrifying micro-organisms of the Eden Estuary, Scotland. In: Cyclic Phenomena in Marine Plants and Animals. (Eds. E. Naylor and R.G. Hartnoll). Pergamon Press, Oxford, pp 259.

Christofi, N., Preston, T. and Stewart, W.D.P. (1981). Endogenous nitrate production in an experimental enclosure during summer stratification. Water Res., 15; 343.

Collins, V.G. (1963). The distribution and ecology of bacteria in freshwater. Proc. Soc. Wat. Treat. Exam., 12; 40.

Dunn, G.M., Herbert, R.A. and Brown, C.M. (1978). Physiology of denitrifying bacteria from tidal mud-flats in the River Tay. In: Physiology and Behaviour of Marine Organisms. (Eds. D.S. McLusky and A.J. Berry). Pergamon, Oxford. pp. 135.

Dunn, G.M. Herbert, R.A. and Brown, C.M. (1979). Influence of oxygen tension on nitrate reduction by a Klebsiella sp. growing in chemostat culture. J. Gen. Microbiol., 112: 379.

Focht, D.D. and Verstraete, W. (1977). Biochemical ecology of nitrification and denitrification. In: Advances in Microbial Ecology. Vol. I. Ed. M. Alexander, Plenum Press, New York. pp. 135.

Herbert, R.A. (1980). Nitrogen cycling, with particular reference to nitrate dissimilation in estuarine and marine sediments. Society for General Microbiology Quarterly, 7: 65.

Iizumi, H., Hattori, A. and McRoy, C.P. (1980). Nitrate and nitrite in interstitial waters of eelgrass beds in relation to the rhizosphere. J. Exp. Mar. Biol. Ecol., 47: 191.

Johnston, D.W., Holding, A.J. and McCluskie. (1974). Preliminary comparative studies on denitrification and methane production in Loch Leven, Kinross and other freshwater lakes. Proc. Roy. Soc. Edin, B., 74: 123.

Jones, J.G. (1971). Studies on freshwater bacteria: factors which influence the population and its activity. J. Ecol., 59: 593.

Jones, J.G. (1979). Microbial nitrate reduction in freshwater sediments. J. Gen. Microbiol., 115: 27.

Jones, J.G. and Simon, B.M. (1981). Differences in microbial decomposition processes in profundal and littoral lake sediments, with particular reference to the nitrogen cycle. J. Gen. Microbiol., 123: 297.

Jones, J.G., Downes, M.T. and Talling, I.B. (1980). The effect of sewage effluent on denitrification in Grasmere (English Lake District). Freshwat. Biol., 10: 341.

Koike, I. and Hattori, A. (1978). Simultaneous determination of nitrification and nitrate reduction in coastal sediments by a ^{15}N dilution technique. Appl. Environ. Microbiol., 35: 853.

Little, J.E., Sjogren, R.E. and Carson, G.R. (1979). Measurement of proteolysis in natural waters. Appl. Environ. Microb., 37: 900.

Mackereth, F.J.H., Heron, J. and Talling, J.F. (1978). Some Revised Methods of Water Analysis for Limnologists. Scient. Publs. Freshwat. Biol. Ass. No. 36.

Meiklejohn, J. (1965). Microbiological studies on large termite mounds. Rhod. Zamb. Mal. J. Agric. Res., 3: 67.

McCaffrey, R.J., Meyers, A.C., Davey, E., Morrison, G., Bender, M., Luedtke, N., Cullen, D., Froelich, P. and Klinkhammer, G. (1980). The relation between pore water chemistry and benthic fluxes of nutrients and manganese in Narragansett Bay, Rhode Island. Limnol. Oceanogr., 25: 31.

McCarthy, J.J. (1980). Nitrogen. In: The Physiological Ecology of Phytoplankton. (Ed. I. Morris). Studies in Ecology. Vol. 7, Blackwells, Oxford. pp. 191.

Owens, N.J.P., Christofi, N. and Stewart, W.D.P. (1979). Primary production and nitrogen cycling in an estuarine environment. In: Cyclic Phenomena in Marine Plants and Animals. (Eds. E. Naylor and R.G. Hartnoll). Pergamon, Oxford. pp. 249.

Painter, H.A. (1977). Microbial transformations of inorganic nitrogen. Prog. Wat. Tech., 8: 3.

Rosenfeld, J.K. (1979). Ammonium adsorption in nearshore anoxic sediments. Limnol. Oceanogr., 24: 356.

Rowe, R., Todd, R. and Waide, J. (1977). Microtechnique for most-probable-number analysis. Appl. Environ. Microbiol. 33: 675.

Solorzano, L. (1969). Determination of ammonia in natural waters by the phenolhypochlorite method. Limnol. Oceanogr. 14: 799.

Sorensen, J. (1978a). Capacity for denitrification and reduction of nitrate to ammonia in a coastal marine sediment. Appl. Environ. Microbiol., 35: 301.

Sorensen, J. (1978b). Denitrification rates in a marine sediment as measured by the acetylene inhibition technique. Appl. Environ. Microbiol., 36: 139.

Sorensen, J. (1978c). Occurrence of nitric and nitrous oxides in a coastal marine sediment. Appl. Environ. Microbiol., 36: 809.

Stewart, W.D.P. (1967). Nitrogen turnover in marine and brackish habitats II. Use of ^{15}N in measuring nitrogen-fixation in the field. Ann. Bot. N.S., 31: 385.

Todd, R.L. and Nuner, J.H. (1973). Comparison of two techniques for assessing denitrification in terrestrial ecosystems. Bull. Ecol. Res. Comm. (Stockholm), 17: 277.

Vanderborght, J.P. and Billen, G. (1975). Vertical distribution of nitrate concentration in interstitial water of marine sediments with nitrification and denitrification. Limnol. Oceanogr., 20: 953.

PARTICIPANTS

ACKROYD, D.R.   Department of Marine Science, Plymouth
                      Polytechnic, PLYMOUTH, Devon, UK.

ALEXANDER, W.R.   Department of Earth Science, Leeds
                      University, LEEDS, UK.

ANDERSON, F.E.   Department of Earth Sciences and Jackson
                      Est. Lab., University of New Hampshire,
                      DURHAM, N.H. USA.

BALE, A.J.   Institute for Marine Environmental Research,
                      Prospect Place, The Hoe, PLYMOUTH, Devon, UK.

CAMERON, F.J.   Department of Microbiology, University of
                      Liverpool, PO Box 174, LIVERPOOL, UK.

COLE, J.A.   Water Research Centre, PO Box 16, Henley Road,
                      Medmenham, MARLOW, Bucks, UK.

CORLETT, J.   President, Estuarine and Brackish Water
                      Sciences Association, Ben Fold, Field Head,
                      Outgate, AMBLESIDE, Cumbria, UK.

CROSBY, S.   Plymouth Polytechnic, PLYMOUTH, Devon, UK.

DAVISON, W.   Freshwater Biological Association, The Ferry
                      House, AMBLESIDE, Cumbria, UK.

DUCK, R.W.   Department of Geology, The University,
                      DUNDEE, UK.

DUFFY, B.   Department of Oceanography, The University,
                      SOUTHAMPTON, UK.

EDWARDS, C.   Department of Microbiology, University of
                      Liverpool, PO Box 174, LIVERPOOL, UK.

HARVEY, B.R.	Ministry of Agriculture, Fisheries and Food, Fisheries Radiobiological Laboratory, Hamilton Dock, LOWESTOFT, Suffolk, UK.
HEAD, P.	North West Water Authority, Head Office, WARRINGTON, Lancashire, UK.
HETHERINGTON, J.A.	Scottish Development Department, Pentland House, EDINBURGH, UK.
HEATON, B.	Biomedical Physics Department, University of Aberdeen, Forester Hill, ABERDEEN, UK.
HERBERT, R.A.	Department of Biological Sciences, University of Dundee, DUNDEE, UK.
HIGGINS, D.C.	Freshwater Biological Association, The Ferry House, AMBLESDIE, Cumbria, UK.
HILTON, J.	Freshwater Biological Association, The Ferry House, AMBLESIDE, Cumbria, UK.
JONES, M.V.	Department of Microbiology, University of Liverpool, PO Box 174, LIVERPOOL, UK.
KENNEDY, H.A.	Department of Earth Sciences, Leeds University, LEEDS, UK.
KERSHAW, P.J.	Ministry of Agriculture, Fisheries and Food, Fisheries Radiobiological Laboratory, Hamilton Dock, LOWESTOFT, Suffolk, UK.
KINSMAN, D.J.J.	Freshwater Biological Association, The Ferry House, AMBLESIDE, Cumbria, UK.
LAXEN, D.	Department of Geology, University of Edinburgh, West Mains Road, EDINBURGH, UK.
LITTLE, D.I.	Field Studies Council, Oil Pollution Research Unit, Orielton Field Centre, PEMBROKE, Dyfed, UK.
McMANUS, J.	Tay Estuary Research Centre, University of Dundee, DUNDEE, UK.
MANTZ, P.A.	Department of Civil Engineering, Imperial College, Exhibition Road, LONDON, UK.
MARSH, J.G.	Department of Marine Science, Plymouth Polytechnic, PLYMOUTH, Devon, UK.

PARTICIPANTS

MILLWARD, G.E.	Department of Marine Science, Plymouth Polytechnic, PLYMOUTH, UK.
MORRIS, A.W.	Institute for Marine Environmental Research, Prospect Place, The Hoe, PLYMOUTH, Devon, UK.
NEALSON, K.H.	Department of Microbiology - SC42, University of Washington, SEATTLE, USA.
OWENS, N.J.P.	Institute for Marine Environmental Research, Prospect Place, The Hoe, PLYMOUTH, Devon, UK.
PARKER, W.R.	Institute of Oceanographic Sciences, Crossway, TAUNTON, Somerset, UK.
READMAN, J.W.	Plymouth Polytechnic, Department of Environmental Sciences, PLYMOUTH, Devon, UK.
ROWLATT, S.	Ministry of Agriculture, Fisheries and Food, Remembrance Avenue, BURNHAM ON CROUCH, Essex, UK.
SHOLKOVITZ, E.	Geology Department, University of Edinburgh, West Mains Road, EDINBURGH, UK.
SILLS, G.S.	Department of Engineering Science, University of Oxford, Parks Road, OXFORD, UK.
TEBO, B.M.	A-008, Scripps Institution of Oceanography, LA JOLLA, California, USA.
THOMAS, A.G.	Institute of Hydrology, WALLINGFORD, Oxon, UK.
TIPPING, E.	Freshwater Biological Association, The Ferry House, AMBLESIDE, Cumbria, UK.
TRUESDALE, V.W.	Institute of Hydrology, WALLINGFORD, Oxon, UK.
TURNER, D.	Marine Biological Association, The Laboratory, Citadel Hill, PLYMOUTH, Devon, UK.
VELTMAN, M.	Municip of Rotterdam, Marconistraat 12, ROTTERDAM, Holland.
WATSON, P.	Institute for Marine Environmental Research, Prospect Place, The Hoe, PLYMOUTH, Devon, UK.
WHALLEY, P.	Newcastle-upon-Tyne University, NEWCASTLE-UPON-TYNE, UK.

WILLIAMS, D.J.A.	Department of Chemical Engineering, University College, SWANSEA, UK.
YATES, J.	Water Research Centre, Stevenage Laboratory, Elder Way, STEVENAGE, Herts, UK.

INDEX

Page numbers underlined indicate Figures;
with suffix 't' entry occurs in a Table.

Adsorption, 11-20
  of humic substances, 32-37
Al:Si ratio
  effect of particle size on,
     101-105, 102, 103, 103t
  of reference minerals, 91t
Americium, in sediments, 127
  in Irish Sea, 136, 137-143
Anoxic sediments, 137-138

Bacteria
  iron oxidising, 197-212, 199t,
     202
    autotrophic, 197-198, 199t
    tests for, 203
    effect of salinity on, 209,
       210
    Gram-negative, 198
    heterotrophic, 198
    in Tamar Estuary, 200-212
      sampling sites, 201
    isolation of, 200-202
    pH range, for growth, 199t
    respiration rate, 203
  nitrogen cycling, 132, 215-226
    enumeration technique, 216
    seasonal variation in, 219,
       220
Bioturbation, 134
Botlekhaven, 174, 176, 177, 179t,
   181, 184t, 185t

Caesium 137, 143-144, 146
  sediment depth profiles, 146,
     148, 149-152, 149, 150,
     151
Cation exchange capacity (CEC),
     1, 2, 12, 17-18, 20, 23,
     28
Charon model, 185-186
Clay/electrolyte interactions,
     1-13
Consolidation, see Sediment
     consolidation
Coulter Counter
  for particle size analysis,
     47-49, 53-73
    data, extrapolation of,
       70-72, 71
Curium, in anoxic sediments,
     137-138

Density profiles, River Parrett
     sediment, 113-118, 114,
     115, 118
Diffusion coefficient, 143-145
Distribution ratio (Kd), 128,
     129t
Donnan model, 1, 3-5, 8, 9, 11
Dredged silt/spoil
  disposal of, 171-172
  contaminants in, 173, 178-180,
     179t, 182

235

Eden Estuary, nitrogen cycling
in, 215-226, 217t, _218_,
_220_, 222t, _223_
Eh *see* Redox potential
Electrokinetic shear plane of
humic substances, 41-44,
_43_
Electrophoresis, 37
Electrophoretic mobility, 37-41
effect of pH on, 37-39, _38_
effect of Ca^{2+} and Mg^{2+} on,
39-40, _40_
effect of phosphate and
silicate on, 40-41, 42t,
43-44
Element X-ray intensity ratios of
reference minerals, 91t
Enterobacter aerogenes, 198
Enteromorpha, production of
ammonia by, 221, 224
Esthwaite Water
adsorption of humics in, 32-34,
33t, _35_
iron oxides in, 37-39, _38_

Flocculation, 110, 113
Flow rate
Tamar River, 210
effect of, on bacterial
distribution, 210-211, 212
water/sediment interface,
119-122, _119_, _120_, _121_
Forth Estuary
analysis of sediment from,
94-95
sources of clay minerals in,
105

Gallionella ferruginea, 198, 199t
Goethite, adsorption of humic
substances by, 32, 33t,
_34_, _35_
Gouy-Chapman model, _6_, 7-9, 11,
26
Grain size
analysis, _50_, 51-53, _52_, _69_, _71_
effect of, on mineral
composition, 99-105
(*see also* Particle size)

Helmholtz model, 1, _6_, 7, 10, 11,
12, 23
Humic substances, 31-37
adsorption of, 32-37, _34_, _35_,
37t, 41-43

Illite, 90t, 91t
compositional variation in,
99-105, _102_, _103_, 103t,
_104_
Interstitial water *see* Pore water
Iron
oxidation, bacterial,
effect of, on respiration
rate, 203, _204_, _205_
effect of salinity on,
203-208, _205_, 206t, _208_,
211-212
oxides
adsorption of humics by,
32-37
concentration of, in sea
water and pore water,
135t
oxidising bacteria, 197-212,
199t
effect of salinity on, _209_,
210
pH range, for growth of, 199t
Redox reactions, 211
in pore waters, 131, _132_
Irish Sea
fission products in, 143-144
sediments, 128-136, 138-139
pore water, 130-139, _132_,
133-136, 135t, 139
transuranic elements in,
128-136, 138, 139
oxidation state of, 128, 129t

Kolmogorov-Smirnov Test
applied to particle size
analysis data, 55-59

Leptothrix spp, 198, 199t

Maasmond, _174_, 176, _177_, 179t,
_181_, 184t, 185t
Manganese, in pore water, 131,
_132_

Mathematical model(s)
  Charon, 185-186
  clay/electrolyte system, 2-10, 4, (*see also* Donnan model; Gouy-Chapman model; Helmholtz model; Stern model)
  particle size distribution, 77-78, 78t
  radionuclide distribution in sediment, 144-146, 152-153
  sedimentation rate, 144-146
Metallogenium spp, 198, 199t
Mica, 90t, 91t, 98
  compositional variation in 99-105, 100, 102, 103, 103t, 104
Mineral(s)
  identification
    by 'template' matching, 89-90, 97
    clay particle, 89-95, 90t, 91t, 92, 93t, 95t, 217, 207t
  reference, element X-ray intensity ratios, 91t
Mississippi Harbor (Rotterdam), 174, 182, 185
Muscovite, 90t, 91t, 92, 98
  compositional variation in, 99-105, 100

Neptunium, 136
  in Irish Sea sediments, 134-137 137t, 138
Nitrogen
  cycling processes, rates of, 221-224, 221t, 222t, 223, 225
  dissolved inorganic, (DIN) concentrations, in Eden Estuary sediment, 216-219, 217t, 218
    in marine sediments, 132-133
    in pore water, 132, 188, 215, 224
  estimation, 216
Notional Interfacial Content (NIC), 10-20, 17, 18-19, 21-28, 21t, 22t, 26

Oostvoorne Meer, 173, 174, 179t
  ammonia levels, 187, 188
  cadmium concentrations, 190
  chlorinity, 190
  erosion, 173, 182
  oxygen levels, 186-188, 187, 191
  salinity of, 182, 185t
  thermal stratification, 186
  turbidity in, 176, 182, 183
  water quality, 188-193
Oxidation state of transuranic elements in Irish Sea, 128, 129t

Parrett, River, sediment
  density profiles, 113-118, 114, 115, 118
  particle size distribution, 112, 113
Particle size
  analysis
    analytical precision of, 56, 58
    Combwich mud, 112
    comparison of methods, 53-73, 54t
    flowchart for, 50, 52
    sample pretreatment, 60-65
      effect of, 63
    using Coulter Counter, 47-49, 53
      combined with sieve, 57-59, 60, 61, 67-68, 69
  distribution, 75-84
    in Tamar Estuary, 75, 80-84, 81, 82
    log-normal, 77-78, 78t
    measurement of, 75-80, 76
    moments method, 53
    normal, 77-78, 78t
    Rosin Rammler, 77-78, 78t, 80-81, 81, 82, 83
    sampling techniques for estimation of, 79
    effect of, on mineral composition, 99-105, 100, 102, 103, 103t, 104
  populations
    mathematical models, for, 77-78, 78t
    (*see also* Grain size)

Penwhirn Reservoir
  adsorption of humics in, 32–33, 40t
  electrophoretic mobility in, 42t
pH
  effect of
    on electrophoretic mobility, 37–39, 38
    on oxidation state of Np, 137t
  levels, in marine environment, 131t
  range, for growth of iron oxidising bacteria, 199t
Plutonium
  in sea water and pore water, 135t
  in sediments, 127
    in Irish Sea, 133
      anoxic, 137–138
      oxidised, 133–136
    in Ravenglass Estuary depth profiles, 146–148, 147
  oxidation states, 128, 129t
Pollutants
  disposal of, 172
  distribution of, in sediments, 109–110, 122, 127, 182, 184t (see also Sediment, contamination in)
Polycyclic aromatic hydrocarbons (PAH), 155–167
  analysis, 159, 160
  concentrations in Tamar Estuary sediments, 157, 161, 165, 166, 167
  distribution, 155–157, 161–164
  sample
    collection, 157, 158
    preparation, 157, 160
  sources, in Tamar area, 164–171
  structure
    compositional, 162, 163, 164
    molecular, 155, 156
Pore water, 109–125
  chemical content, 127–139, 176, 178–180, 179t, 182, 188
  extraction, sampling technique, 130
    for transuranic analysis, 133

Pore water (continued)
  ferrous iron concentration in, 135t, 188
  in Irish Sea sediments, 130–139, 132, 133–136, 135t, 139
  ph distribution in, 132, 135t
  movement, 109–110, 117–125
  nitrogen, dissolved inorganic, in, 215–219, 217t, 218, 224
    estimation of, 216
  phosphate concentration in, 177, 188
  plutonium in, 134–136, 135t
  pressure, 110–112, 115
  radionuclide concentration in, 144–146
  volume, 119–124, 120, 121, 123

Radionuclides
  in estuarine sediments, 143–153
  in marine sediments, 127–139
Ravenglass Estuary, radionuclide distribution data, 146, 147, 148, 149, 150, 151
Redox potential (Eh), 128–134, 131t, 132, 135t, 211
  effect of, on oxidation state of Np, 137t
Reference minerals
  for electron microanalysis, 89–90, 90t
  X-ray intensity ratios of, 91t
River flow rate, Tamar River, 210
  effect of, on bacterial distribution, 210–211, 212
Rostherne Mere
  adsorption of humics in, 32–33, 33t
Rotterdam harbor
  dredged material from, 171–172

Salinity
  effect of
    on distribution of iron oxidising bacteria, 209, 210
    on NIC, 23–25, 24t
    of Oostvoornse Meer, 182, 185t,

INDEX

Sample preparation
　for particle size analysis, 47, 49-53
　for transmission electron microscope, 89
Sample splitters, 61
Schofield Equation, 21, 25-34
Sediment
　analysis
　　multi-grain, 96-99
　　single-grain, 88-95
　anoxic, 137-138
　contamination in, 176, 178, 179t (see also Pollutants, distribution of, in sediments)
　consolidation, 109-125, 176, 178, 180, 181
　　laboratory simulation, method, 110-113, 111
　　results, 113-123, 114, 115, 117, 118, 119, 120, 121, 123
　state, definitions of, 116
　suspended, settlement of, 110-118
　/water interface, 109-110, 116, 117, 119, 119, 120, 123-124, 145
Sedimentation rate, 144-153, 147, 148, 149, 151
　model, 144-146
Serratia indica, 198
Sodium chloride, effect of, on bacterial iron oxidation, 203-208, 205, 206t, 208
Sphaerotilus natans, 198
Stern model, 5-7, 6, 9, 11
Surface-excess,
　chloride, 25, 27
　ion, 11, 12-13, 17, 36

Tamar Estuary
　autotrophic iron oxidising bacteria in, 200-212
　　effect of salinity on, 205, 206t, 207-208, 208, 209
　clay minerals in, 207, 207t
　　effect of, on bacterial iron oxidation, 207-210, 208
　particle size distribution in, 75, 80-84, 81, 82

Tamar Estuary (continued)
　polycyclic aromatic hydrocarbons in, 157, 161-167
　　collection of, 157, 158
　turbidity distribution in, 80-83, 81, 82
Thiobacillus ferrooxidans, 198, 199t, 207
Transmission electron microscope (TEM), 88-89
　for mineral identification, 88-95, 90t, 91t, 92, 93t, 95t, 98, 105-106
　preparation of samples for, 89, 94
Transuranic elements in marine sediments, 127-139
Turbidity distribution
　in Oostvoornse Meer, 182, 183
　in Tamar estuary, 80-83, 81, 82

Ultrasonic probe, use of, for sample dispersion, 64-65, 65

Void ratio, 119-120, 122

Waalhaven, 174, 176, 177, 178, 179t, 181, 184t, 185t
Water quality
　improvement, 188-193, 190
　　by flushing, 187, 190
　factors influencing, 175
　Oostvoornse Meer, 171-193
Water/sediment interface see Sediment/water interface
Wilcoxon Matched Pairs Test applied to particle size analysis data, 55-59
Windscale, 134, 135t, 143
　effluent, 128, 136, 146, 152

X-ray
　absorption, for settlement measurement, 111-112, 113
　diffraction for mineral identification, 87, 94, 97, 99, 106
　intensity ratios of reference minerals, 91t